T0205503

Highly Integrated Gate Drivers for Si and GaN Power Transistors

Achim Seidel • Bernhard Wicht

Highly Integrated Gate Drivers for Si and GaN Power Transistors

 Springer

Achim Seidel
Robert Bosch GmbH
Reutlingen, Germany

Bernhard Wicht
Leibniz University Hannover
Hannover, Germany

ISBN 978-3-030-68942-1   ISBN 978-3-030-68940-7   (eBook)
https://doi.org/10.1007/978-3-030-68940-7

This Springer imprint is published by the registered company Springer Nature Switzerland AG
The registered company address is: Gewerbestrasse 11, 6330 Cham, Switzerland

Preface

This book explores integrated gate drivers with emphasis on new GaN power transistors, which offer fast switching along with minimum switching losses. In this environment, the gate driver is a key component that enables advanced, energy-efficient, robust, and compact power electronics. As GaN transistors are relatively new, there is still a large innovation potential for suitable integrated gate drivers. IC technology enables miniaturization by scaling down length of interconnections at the power stage. As a main advantage, this reduces parasitic inductances, which allow for even faster switching such that the potential of GaN as well as of conventional Si-based switches can be utilized. Nevertheless, there are still a large number of challenges in the area of gate drivers.

The book investigates solutions on system and circuit level for highly integrated gate drivers. It includes (1) miniaturization by higher integration of subfunctions onto the IC (buffer capacitors) and (2) more efficient switching by a multilevel approach, which also improves robustness in case of extremely fast switching transitions. It presents (3) a concept for robust operation in the highly relevant case that the gate driver is placed in distance to the power transistor. All results are widely applicable.

Fast switching gate drivers require that the driver is capable of high output currents. Depending on the semiconductor technology, conventional gate drivers suffer from large die area occupied by the gate driver output stage resulting in high costs. This book introduces two concepts with various benefits such as area savings and increased driving speed: high-voltage charge storing (HVCS) and high-voltage energy storing (HVES). As a key idea of HVCS, the main bootstrap capacitor is supported by a second bootstrap capacitor that is charged to a higher voltage and ensures high charge allocation when the driver turns on. HVCS achieves ~70 % area reduction of the bootstrap capacitor that is required to provide the charge for the pre-driver pull-up NMOS transistor of the driver output stage. Besides drivers, the proposed bootstrap circuit can also be directly applied to power stages to achieve fully integrated switched mode power supplies or class D output stages. With the current trend towards monolithic integration of gate drivers on the same die as the GaN transistor, HVCS is very attractive to reach the integration of the buffer capacitor, achieving higher reliability, easier usage, and faster switching.

In addition to HVCS, HVES is based on a resonant behavior, basically due to an integrated inductor in the gate charge path. This enables to deliver high-current pulses based on a resonant discharging of a high-voltage buffer capacitor over this inductor to the gate. The key idea is to transfer the energy stored in the buffer capacitor to the gate, charging it, e.g., to 5 V. Since the energy in the capacitor is proportional to the square of its initial voltage (e.g., 15 V), the circuit can provide a large amount of charge on a small die area. Moreover, HVES shows a ~25 % higher gate drive speed compared to a conventional gate driver, even at a three times larger gate loop inductance. Hence, HVES is in particular suitable for two scenarios: (1) if the inductance of the gate loop is a limiting factor for higher gate driver speed, and (2) it enables the integration of buffer capacitors capable of driving even large types of GaN transistors at an appropriate gate driver efficiency. An experimental implementation delivers up to 11 nC gate charge from a fully integrated buffer capacitor (no external components) along with a robust bipolar and three-level gate drive scheme.

Based on high-voltage energy storing, a gate driver concept and various implementation options are proposed that allow for placing the gate driver in large distance from the power transistor without sacrificing the switching performance. Experimental results show an increase in the gate drive speed of up to ~40 %. Transient measurements show $60 \, \mathrm{V \, ns^{-1}}$ steep voltage transitions at a GaN transistor, with a distance of the gate driver to the power transistor of ~25 cm. This concept allows for different temperatures and different substrates for the gate driver and transistor, resulting in better component placement depending on application constraints.

The objective of this book is to provide a systematic and comprehensive insight into gate drive techniques for fast switching power stages. Both theoretical and practical aspects are covered. Design guidelines are derived for various gate drivers, level shifters, and supporting circuits as well as for the on-chip buffer capacitors in case of HVCS and HVES. The material will be interesting for design engineers in industry as well as researchers who want to learn about and implement gate drivers.

This book is based on our research at Reutlingen University, Reutlingen, Germany, and at the Institute for Microelectronic Systems at Leibniz University Hannover, Hannover, Germany. The work was partially sponsored by the State of Baden-Württemberg and by the Federal Ministry of Education and Research Germany (Grand No. 16ES0080). We are grateful to all team members for the invaluable support as well as the enjoyable collaboration. A special thanks goes to our families, without their love and support this book would not have been possible.

Filderstadt, Germany   Achim Seidel

Gehrden, Germany   Bernhard Wicht
December 2020

Contents

Acronyms

List of Abbreviations

2DEG	Two-dimensional electron gas
Al	Aluminum
BCD	Bipolar complementary metal oxide semiconductor (CMOS) double-diffused metal oxide semiconductor (DMOS)
BJT	Bipolar junction transistor
CMOS	Complementary metal oxide semiconductor
CMTI	Common-mode transient immunity
DMOS	Double-diffused metal oxide semiconductor
EMC	Electromagnetic compatibility
EMI	Electromagnetic interference
FOM	Figure of merit
GaN	Gallium nitride
GIT	Gate injection transistor
HVCS	High-voltage charge storing
HVES	High-voltage energy storing
IC	Integrated circuit
IGBT	Insulated gate bipolar transistor
LDO	Linear regulator
LiDAR	Light detection and ranging
MOS	Metal oxide semiconductor
MOSFET	Metal oxide semiconductor field effect transistor
NMOS	N-channel metal oxide semiconductor
PCB	Printed circuit board
PFC	Power-factor correction
PMOS	P-channel metal oxide semiconductor
PWM	Pulse-width modulation
QFN	Quad-flat no-lead package
Si	Silicon

SiC		Silicon carbide
SJMOSFET		Superjunction MOSFET
SMD		Surface-mounted device
SOA		Safe-operating area
SOIC		Small-outline integrated circuit

List of Symbols

A	m^2	Area of the electrodes of a plate capacitor
B_p		Buffer in gate driver circuit for driving large gate loops
C_{B1}	F	Low-voltage bootstrap capacitor of the HVCS circuit
C_{B2}	F	High-voltage bootstrap capacitor of the HVCS circuit
C_{BST}	F	Bootstrap capacitor
C_{Buf}	F	Buffer capacitor (called C_{HVp} / C_{HVn} in the implemented gate driver circuits)
C_{cpl}	F	Coupling capacitance between the gate driver low-side and high-side
C_{CP}	F	Capacitor of the "V_{DRV} C_P" circuit
C_{DS}	F	Drain-source capacitance
C_{DRV}	F	Gate driver buffer capacitor
C_G	F	Total gate capacitance of the power transistor
$C_{G,eq}$	F	Equivalent linear gate capacitance
C_{GD}	F	Gate-drain capacitance
C_{GS}	F	Gate-source capacitance
$C_{GS,add}$	F	Capacitor connected in parallel to the gate-source terminals of the power transistor
C_{HVn}	F	High-voltage capacitor of the HVES circuit (transistor's source side)
C_{HVp}	F	High-voltage capacitor of the HVES circuit (transistor's gate side)
C_{HVx}	F	High-voltage capacitors of the HVES circuit (transistor's gate/source side)
CLK_{CP}	F	Clock signal of the "V_{DRV} C_P"
C_{loop}	F	Equivalent gate loop capacitance
C_{oss}	F	Output capacitance of the power transistor ($C_{GD} + C_{DS}$)
$C_{o(ER)}$	F	Effective output capacitance of the power transistor—energy related
$C_{o(TR)}$	F	Effective output capacitance of the power transistor—time related

C_P	F	Charge pump pumping capacitor
C_{plate}	F	Capacitance of a plate capacitor
d	m	Distance between the electrodes of a plate capacitor
D		Damping factor of the gate loop
D_B		Bootstrap diode of high-side gate driver supply
D_{B1}		Bootstrap diode for low-voltage supply rail of the HVCS circuit
D_{B2}		Bootstrap diode for high-voltage supply rail of the HVCS circuit
D_{C1}		Diode of the charge pump circuit
D_{C2}		Diode of the charge pump circuit
D_{clpn}		Clamping diode of the gate driver for large gate loops (transistor's source side)
D_{clpp}		Clamping diode of the gate driver for large gate loops (transistor's gate side)
D_{CP}		Diode of the charge pump circuit
D_{clp}		Clamping diode, protecting the supply-voltage input of a high-side gate driver
D_{FW}		Free-wheeling diode
D_{FWn}		Free-wheeling diode of the gate driver for large gate loops (transistor's source side)
D_{FWp}		Free-wheeling diode of the gate driver for large gate loops (transistor's gate side)
D_R		Rectifier diode of HVES circuit
D_{Rn}		Rectifier diode of HVES circuit (transistor's source side)
D_{Rp}		Rectifier diode of HVES circuit (transistor's gate side)
DRV_{IN}	V	Gate driver control signal
DRV_{INn}	V	Gate driver control signal (transistor's source side)
DRV_{INp}	V	Gate driver control signal (transistor's gate side)
ΔV_{GS}	V	Amplitude of the voltage dip of V_{GS}, caused by the Miller coupling
ϵ_r	$F\,m^{-1}$	Relative permittivity of a dielectric medium
ϵ_o		Electric constant
η		Efficiency of a turn-on event of the HVES / HVCS circuit
η_{HVES}		Efficiency of a turn-on event of the HVES circuit
E_{CBuf}	J	Energy in C_{Buf}
E_{CG}	J	Energy in C_G
E_{in}	J	Energy from C_{Buf} at driver turn-on
$E_{chargingLossCoss}$	J	Dissipated energy when charging C_{oss} from 0 V to the stated V_{DS} via an energy-dissipating element

E_{oss}	J	Energy stored in the fully charged C_{oss}
E_{loss}	J	Energy dissipated during a driver turn-on event
$E_{loss,HVES}$	J	Energy dissipated during a driver turn-on event in the HVES circuit
E_{out}	J	Energy stored in C_G after driver turn-on
E_n	J	Energy from HVES circuit at driver turn-off
E_p	J	Energy from HVES circuit at driver turn-on
f_{CP}	Hz	Clock frequency of the "V_{DRV} C_P"
I_{CP}	A	Load current of the "V_{DRV} C_P" circuit
I_D	A	Drain current of the power transistor
I_L	A	Load current
I_n		Inverter in the gate driver circuits for driving large gate loops
I_{peak}	A	Peak driver output current
I_G	A	Gate drive current
$I_{G,sink}$	A	Sink gate drive current
$I_{G,sink,max}$	A	Maximum sink gate drive current
$I_{G,source}$	A	Source gate drive current
$I_{G,source,max}$	A	Maximum source gate drive current
k_1		Scaling factor of C_{B1} accounting for process tolerances
k_2		Scaling factor of C_{B2} accounting for process tolerances
$k_{1,max}$		Maximum specified scaling factor of C_{B1} accounting for process tolerances
$k_{2,max}$		Maximum specified scaling factor of C_{B2} accounting for process tolerances
$k_{1,min}$		Minimum specified scaling factor of C_{B1} accounting for process tolerances
$k_{2,min}$		Minimum specified scaling factor of C_{B2} accounting for process tolerances
L	H	Inductor as part of the equivalent HVES circuit (called L_{HVp} / L_{HVn} in the implemented gate driver circuits)
L_{HV}	H	Inductor as part of the HVES circuit
L_{HVn}	H	Inductor of the HVES circuit (transistor's source side)
L_{HVp}	H	Inductor of the HVES circuit (transistor's gate side)
L_{HVx}	H	Inductor of the HVES circuit (transistor's gate/source side)
L_{loop}	H	Gate loop inductance
L_{par}	H	Parasitic inductance
$LSin$	V	Level-shifter input signal
$LSout$	V	Level-shifter output signal

M_{CP}		Low-voltage transistor of the "CHV C_P" circuit
M_{CP2}		High-voltage transistor of the "V_{DRV} C_P" circuit
M_{Hn}		High-voltage transistor of the gate driver for large gate loops (transistor's source side)
M_{HVn}		High-voltage transistor of the gate driver based on HVES (transistor's source side)
M_{Hp}		High-voltage transistor of the gate driver for large gate loops (transistor's gate side)
M_{HVp}		High-voltage transistor of the gate driver based on HVES (transistor's gate side)
M_{HVx}		High-voltage transistor M_{HVp} or M_{HVn}
M_{Lp}		High-voltage transistor of the gate driver for large gate loops (transistor's gate side)
M_{Ln}		High-voltage transistor of the gate driver for large gate loops (transistor's source side)
$MN1_{off}$	V	Signal in the level-down shifter circuit in the gate driver based on HVCS
$MN2_{EN}$	V	Signal in the gate driver based on HVCS
MN_{B1}		High-voltage transistor of the "V_{HV} control" circuit
MN_{B2}		High-voltage transistor of the "V_{HV} control" circuit
MN_R		Transistor for active rectification in the gate driver based on HVES
MN_n		PMOS transistor of the gate driver based on HVES (transistor's source side)
MN_p		PMOS transistor of the gate driver based on HVES (transistor's gate side)
MN_n		NMOS transistor of the gate driver based on HVES (transistor's source side)
MN_p		NMOS transistor of the gate driver based on HVES (transistor's gate side)
ω_o	1/s	Resonance frequency of the HVES circuit
R_{Bn}	Ω	Resistor of the gate driver for large gate loops based on HVES (transistor's source side)
R_{Bp}	Ω	Resistor of the gate driver for large gate loops based on HVES (transistor's gate side)
$R_{DS,on}$	Ω	Transistor on-state resistance
R_G	Ω	Gate resistor
$R_{G,on}$	Ω	Turn-on gate resistor
$R_{G,off}$	Ω	Turn-off gate resistor
R_{loop}	Ω	Gate loop resistance
R_{par}	Ω	Parasitic gate loop resistance
S	V	Source potential of the power transistor

Q_{Coss}	A s	Charge required to fully charge C_{oss} during transistor off-state
Q		Quality factor of the gate loop
Q_G	A s	Gate charge of the driven transistor
Q_{GD}	A s	Gate-drain charge of the driven transistor at specified drain-source switching voltage
Q_{Miller}	A s	Charge coupling into the gate caused by the Miller effect
$Q_{G,ideal}$	A s	Delivered gate charge of the ideal HVES circuit
Q_{ov}	A s	Excessive gate charge Q_G
Q_{max}	A s	Maximum total charge that must be delivered by the HVCS circuit
Q_{min}	A s	Minimum total charge that must be delivered by the HVCS circuit
Q_{tot}	A s	Total charge that must be delivered by the HVCS circuit
t	s	Time
t_D	s	Dead time
V_+	V	Supply voltage of high-voltage measurement setup
V_1	V	Low-voltage supply for the HVCS circuit
V_3	V	High-voltage supply for the HVCS circuit
V_{B1}	V	Floating low-voltage supply rail of the HVCS circuit
V_{B2}	V	Floating high-voltage supply rail of the HVCS circuit
V_C	V	Voltage across the buffer capacitor C_{Buf} (called V_{HVp} / V_{HVn} in the implemented gate driver circuits)
V_{CB1}	V	Voltage across the capacitor C_{B1}
V_{CB2}	V	Voltage across the capacitor C_{B2}
V_{CB1o}	V	Initial voltage across the capacitor C_{B1}
V_{CB2o}	V	Initial voltage across the capacitor C_{B2}
$V_{CB1o,min}$	V	Minimum specified initial voltage across the capacitor C_{B1}
$V_{CB2o,min}$	V	Minimum specified initial voltage across the capacitor C_{B2}
$V_{CB1o,max}$	V	Maximum specified initial voltage across the capacitor C_{B1}
$V_{CB2o,max}$	V	Maximum specified initial voltage across the capacitor C_{B2}
$V_{C,min}$	V	Minimum voltage across the capacitor C_{Buf} after driver turn-on event
V_{CP}	V	Voltage at the top plate of the charge pump capacitor C_P

ΔV_1	V	Voltage dip at the capacitor C_{B1}
ΔV_2	V	Voltage dip at capacitor C_{B2}
$\Delta V_{1,max}$	V	Maximum specified voltage dip at capacitor C_{B1}
$\Delta V_{2,max}$	V	Maximum specified voltage dip at capacitor C_{B2}
ΔV_{meas}	V	Measured voltage dip at capacitor C_{B1}
$\Delta V_{1,min}$	V	Minimum specified voltage dip at capacitor C_{B1}
$\Delta V_{2,min}$	V	Minimum specified voltage dip at capacitor C_{B2}
$\Delta V_{2,max}$	V	Maximum specified voltage dip at capacitor C_{B2}
$\Delta V_{1,nom}$	V	Nominal voltage dip at capacitor C_{B1}
V_{DCP}	V	Forward voltage of the diode D_{CP}
V_{DD5}	V	5 V supply rail
V_{DD5V7}	V	$\sim$5.7 V supply rail
V_{DD8}	V	8 V supply rail
V_{DR}	V	Forward voltage of the diode D_R
V_{DRV}	V	Gate driver supply voltage
V_{DS}	V	Drain-source voltage of the power transistor
V_F	V	Diode forward voltage
V_L	V	Voltage across L
V_{5f}	V	Low-voltage floating high-side supply
$V_{F,DB1}$	V	Forward voltage of the diode D_{B1}
$V_{F,DB2}$	V	Forward voltage of the diode D_{B2}
V_G	V	Gate voltage of the driven transistor
V_{G12}	V	Gate voltage of the transistors MN_{B1}/MN_{B2}
V_{GS}	V	Gate-source voltage of the driven transistor
$V_{GS,max}$	V	Maximum rated driver output voltage
V_{GSon}	V	Nominal V_{GS} of power transistor in the on-state
V_{HV}	V	Voltage across the high-voltage capacitors C_{HVx}
V_{HVn}	V	Voltage across the high-voltage capacitor C_{HVn}
V_{HVp}	V	Voltage across the high-voltage capacitor C_{HVp}
V_{HVx}	V	Voltage across the high-voltage capacitors C_{HVp}/C_{HVn}
V_{off}	V	Voltage node in the level shifter circuit
V_{on}	V	Voltage node in the level shifter circuit
V_n	V	Driver output voltage of gate driver for large gate loops (transistor's source side)
V_p	V	Driver output voltage of gate driver for large gate loops (transistor's gate side)
V_{Rloop}	V	Voltage across the resistance R_{loop}
V_{SUP}	V	Gate driver supply voltage
V_{SW}	V	Switching voltage node
V_{th}	V	Gate-threshold voltage
V_{xoff}	V	Detection node in the level shifter circuit
V_{xon}	V	Detection node in the level shifter circuit

Chapter 1
Introduction

1.1 Motivation

System integration is an ongoing trend for power electronics and a main task for the
next decade [1] in growth areas like renewable energy, e-mobility or industry, and
others, as illustrated in Fig. 1.1. Figure 1.2 demonstrates the continuously increasing
power densities of power electronics applications over years by means of 300 W
voltage converters [1, 2].

A key approach for reducing the system size is increasing the switching
frequency. While this is very effective in shrinking down the size of passives, higher
dynamic losses lead to more cooling effort. This counteracts the volume reduction,
because large heat sinks and other cooling mechanisms need to be installed [3].
Therefore, it is crucial to minimize losses.

Figure 1.2 indicates the approximate evolution of various power transistor types
referred to the beginning of their commercialization. Until the 1970s, BJTs were
used as power switches. However, they were not suitable for various emerging power
electronics applications [4]. In the mid-1970s, power metal oxide semiconductor
field effect transistors (MOSFETs) entered the market with superior switching
characteristics [4]. The IGBT came up in 1985 with a higher voltage and current
capability, combined with an isolated gate structure like the power MOSFET [5].
In 1998, the SJMOSFET was commercialized, which can switch significantly
faster than previous devices and exhibits lower forward resistance compared to
conventional power MOSFETs at high breakdown voltages [8]. In recent years,
wide-bandgap materials, like GaN and SiC, came into focus for power transistors
because of inherently better characteristics leading to low conduction and dynamic
losses, high temperature capability, and better thermal behavior [12]. Especially
in the segment of 600 V power devices, GaN transistors take over market share
from silicon transistors [13]. Each new transistor type represents major progress,
significantly reducing losses compared to previous devices. Reducing parasitic
device capacitances at a certain on-state resistance leads to lower dynamic losses,

© The Author(s), under exclusive license to Springer Nature Switzerland AG 2021
A. Seidel, B. Wicht, *Highly Integrated Gate Drivers for Si and GaN Power
Transistors*, https://doi.org/10.1007/978-3-030-68940-7_1

Fig. 1.1 Power electronics applications

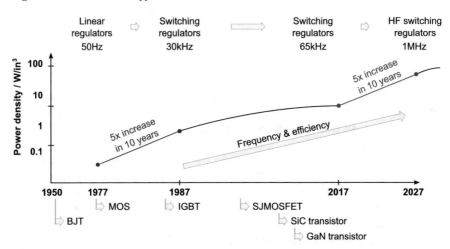

Fig. 1.2 Power density of 300 W AC-DC voltage converters over time, based on [2]. It also shows the evolution of the various power devices (bipolar junction transistor (BJT) [4], metal oxide semiconductor (MOS) [4, 5], insulated gate bipolar transistor (IGBT) [5–7], superjunction MOSFET (SJMOSFET) [5, 8], silicon carbide (SiC) [9], gallium nitride (GaN) [10, 11])

related to faster switching transitions with higher dV/dt and to lower gate driver losses. Even the well-established IGBT, which seems to be at its technological limit, improves in performance due to new designs and new materials. Zong et al. [14] shows market and technology trends giving a promising forecast for IGBTs. There is an ongoing trend towards improving SJMOSFETs [15, 16]. SiC and especially GaN transistors are improving rapidly in technology, because they are in the focus of research and have not been on the market for long.

Modern power transistors demand for gate drivers with a high gate charging speed to perform fast switching transitions, reducing the dynamic losses. Figure 1.3 shows a generic conventional gate driver circuit. For transistor turn-on, charge is brought onto the gate and for turn-off pulled out of the gate again. It requires a buffer capacitor C_{DRV}, which is also called stabilization capacitor, to buffer the

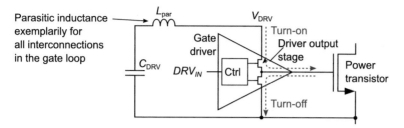

Fig. 1.3 Gate driver and power transistor

drive voltage at turn-on. With values of typically >10 nC, C_{DRV} can usually not be implemented on-chip. However, in case of fast switching, the effectiveness of C_{DRV} is severely reduced, as there is always some parasitic inductance L_{par} from interconnects, bond wires, etc. in series. L_{par} limits the switching performance at driver turn-on. To prevent a gate overshoot or ringing at V_{GS}, conventional solutions place a damping resistor slowing down the driver turn-on event. To keep the inductance small, effort has to be spent on assembly and layout design [17]. This is especially the case on ceramic substrates with only a single wiring layer, which is an attractive solution for GaN transistors [18]. Reducing the influence of parasitic inductances in the gate loop is one reason that pushes the trend towards higher integration levels of gate drivers [19, 20]. Especially, in applications with a large number of power switches, gate drivers contribute significantly to the circuit board volume [21–23]. Highly integrated GaN half-bridges including the gate driver circuit in a single package or even on the same die are on the rise [24–33]. In contrast to discrete solutions, integrated gate drivers increase reliability and functionality, reduce the component count, and achieve smaller volume and lower cost [34].

High-speed gate drivers have a high output-current capability and occupy large die areas, mainly determined by the gate driver output stage (Fig. 1.3). Bootstrap capacitors buffer the charge of the pre-driver driving the pull-up transistor of a gate driver output stage comprising two n-channel metal oxide semiconductors (NMOS). After gate charge delivery, a voltage drop ΔV at the bootstrap capacitor occurs. The allowed voltage drop needs to be small to ensure a proper circuit functionality. Conventionally, the bootstrap capacitor requires a large die area due to the relation $Q = C \cdot \Delta V$. Hence, the area-efficient integration of buffer capacitors means a cost-effective implementation and is of high interest.

Gate drivers leverage the potential of power transistors by optimized and reliable switching at the limits of their safe-operating area (SOA) and electromagnetic interference (EMI) requirements. Drivers can be adapted to the special character-istics of the power transistor like, for example, in a three voltage-level gate drive scheme for GaN transistors [35]. However, this requires additional components like buffer capacitors that are typically not suitable for integration. Another restriction to operate the power device at its limit comes from the different requirements for the optimal operation of gate driver and power transistor. Power transistors can conduct high currents, so ceramic substrates, for example, can be preferred to achieve better thermal performance and higher current capability [18]. Furthermore, they can operate at high temperatures, require cooling attachments, and optimized placement in the application. However, gate drivers do often physically not fit in these environments. Thermal constraints block the optimal placement of gate drivers. Determined by application, there might not be enough space for the gate driver close to the power switch. GaN transistors have naturally a higher temperature stability. However, one reason why this potential cannot be fully retrieved is the gate driver that is typically made in a silicon technology, which cannot operate at these high temperatures. This results in larger gate loops or trade-offs regarding optimal power transistor and gate driver environments, limiting the switching performance.

Hence, gate drivers play an important role in building compact and cost-efficient power electronics. This is supported by gate drivers, which reach for a high degree of IC-level integration, provide efficient and fast switching, support a great placement flexibility, and, hence, leverage the potential of the power transistors.

1.2   Scope and Outline of This Book

An overview of the scope of this work is shown in Fig. 1.4. The main trend goes towards compact power electronics, while cost-efficiency is an important parameter. Three goals for gate drivers are derived and addressed in this book as follows:

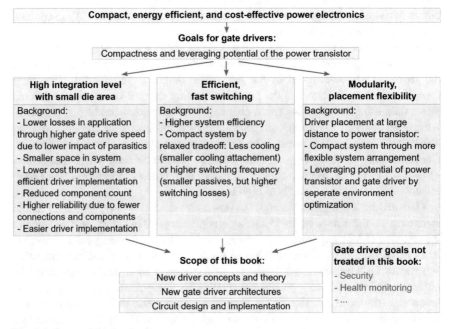

Fig. 1.4 Scope of this book

1. IC-level integration enables a compact gate control, minimizing the effect of parasitics resulting in a fast and robust switching. In particular, the area-efficient integration of buffer capacitors is investigated in this work.
2. For achieving efficient and fast switching, the scope is set on investigations of the gate loop together with gate drivers, providing high gate charge speed, and architectures supporting various gate drive schemes.
3. Flexible gate driver placement is addressed by gate drivers with high gate drive speed even if they are placed far away from the power transistor, enabling separate optimization of the gate driver and power transistor environments.

To achieve the requirements for the gate drivers, the focus of this book is on (1) new gate drive concepts including theory and calculation guidelines, (2) new gate driver architectures for applying the new gate drive concepts, and (3) circuit design solutions, design aspects, and experimental verification of the implemented gate drivers. The gate drivers are designed for applications in the sub-10 kW-range with voltages ≤ 400 V and currents ≤ 10 A. This is a typical range covered by GaN transistors for applications mentioned in Sect. 2.2.3. Nevertheless, the presented solutions are widely adaptable to various applications.

The outline of this book is as follows. Chapter 2 gives fundamentals about gate drivers, silicon and GaN power transistors, and application. Chapter 3 describes a gate driver for silicon transistors. Section 3.2 presents three options of an area-efficient bootstrap circuit based on the concept of high-voltage charge storing (HVCS) [36], introduced in Sect. 3.1. Section 3.4 elaborates various further applications for the proposed bootstrap circuit. In Sect. 3.5, a sizing guideline for the required bootstrap capacitors is given. Experimental results in Sect. 3.6 verify the concept and the calculations of Sect. 3.5. Chapter 4 presents gate drivers for GaN transistors based on the concept of high-voltage energy storing (HVES) [20]. The concept and theory of HVES introduced in Sect. 4.1 is compared to other capacitor implementation methods. The gate driver implementations are described in Sect. 4.2 and experimentally verified, shown in Sect. 4.3. Section 4.4 discusses limitations of drivers based on HVES versus the state of the art. It also shows application fields in comparison to drivers based on other capacitor implementation methods. Chapter 5 proposes a gate driver designed for a large distance to the power transistor [37]. The concept is described in Sect. 5.2, the implementation in Sect. 5.3, and experimental results with a comparison to prior art in Sect. 5.4. Chapter 6 gives an outlook for further developments and research topics, and Chap. 7 finally concludes the results of this work.

References

1. *ECPE Position Paper on Next Generation Power Electronics Based on Wide Bandgap Devices – Challenges and Opportunities for Europe* (2016, May). European Center for Power Electronics e.V.
2. Oliver, S., Xue, L., & Huang, P. (2017, December). From science fiction to industry fact: GaN power ICs enable the new revolution in power electronics. In *Bodo's Power Systems* (pp. 18–22).

3. März, M., Schletz, A., Eckardt, B., Egelkraut, S., & Rauh, H. (2010, March). Power Electronics system integration for electric and hybrid vehicles. In *Proceedings of the 6th International Conference Integrated Power Electronics Systems* (pp. 1–10).
4. Saxena, R. S., & Kumar, M. J. (2012). *Trench gate power MOSFET: Recent advances and innovations*. arXiv preprint arXiv:1208.5553.
5. Butler, S. W. (2019). Enabling a powerful decade of changes. *IEEE Power Electronics Magazine, 6*(1), 18–20.
6. Iwamuro, N., & Laska, T. (2017). IGBT history state-of-the-art, and future prospects. In *IEEE Transactions on Electron Devices, 64*(3), 741–752. ISSN: 0018-9383. https://doi.org/10.1109/TED.2017.2654599
7. Iwamuro, N., & Laska, T. (2018). Correction to "IGBT history, state-of-the-art, and future prospects" [Mar. 17 741-752]. *IEEE Transactions on Electron Devices, 65*(6), 2675. ISSN: 0018-9383. https://doi.org/10.1109/TED.2018.2821172
8. Udrea, F., Deboy, G., & Fujihira, T. (2017). Superjunction power devices, history, development, and future prospects. *IEEE Transactions on Electron Devices, 64*(3), 713–727. ISSN: 0018-9383. https://doi.org/10.1109/TED.2017.2658344
9. Rabkowski, J., Peftitsis, D., & Nee, H. (2012). Silicon carbide power transistors: A new era in power electronics is initiated. *IEEE Industrial Electronics Magazine, 6*(2), 17–26. ISSN: 1932-4529. https://doi.org/10.1109/MIE.2012.2193291
10. Incorporation, GaN Systems. (2017). *Gallium Nitride Power Transistors in the EV World*. GaN Systems Incorporation. https://gansystems.com/wp-content/uploads/2018/01/The-Benefits-of-Gallium-Nitride-Power-Transistors-Span-Multiple-Markets.pdf
11. Ma, Y. (2011). *EPC eGan® FETs Transistor Application Readiness: Phase Four Testing*. Efficient Power Conversion Corporation. https://epc-co.com/epc/Portals/0/epc/documents/product-training/EPC_Phase_Four_Rel_Report.pdf
12. Kaminski, N. (2009, September). State of the art and the future of wide band-gap devices. In *Proceedings of the 13th European Conference on Power Electronics and Applications* (pp. 1–9).
13. Barbarini, E., & Le Troadec, V. (2016, March). *GaN on Si HEMT vs SJ MOSFET: Technology and Cost Comparison – Will SJ MOSFETs still be attractive compared to GaN devices?* Yole Développment. https://www.i-micronews.com/images/Flyers/Power/Yole_GaN_on_Si_HEMT_vs_SJ_MOSFET_technology_and_cost_comparison_March_2016.pdf
14. Zong, Z., Villamor, A., Liao, J., & Lin, H. (2017, August). *IGBT Market and Technology Trends 2017*. Yole Développment. https://www.i-micronews.com/category-listing/product/igbt-market-and-technology-trends-2017.html?utm_source=PR&utm_medium=email&utm_campaign=IGBT_MarketStatus_YOLE_August2017
15. Zhu, Z., Wang, F., & Shen, J. K. (2018, March). 7um pitch deep trench super junction process development. In *Proceedings of China Semiconductor Technology International Conference (CSTIC)* (pp. 1–3). https://doi.org/10.1109/CSTIC.2018.8369169
16. Hancock, J., Stueckler, F., & Vecino, E. (2013, April). *Cool MOS^{TM} C7: Mastering the art of quickness a technology description and design guide*. Infineon Technologies AG.
17. Moroney, M. (2015). *New power switch technology and the changing landscape for isolated gate drivers*. Norwood: Analog Devices, Inc.
18. Yu, C., Buttay, C., & Labouré, É. (2017, February). Thermal management and electromagnetic analysis for GaN devices packaging on DBC substrate. *IEEE Transactions on Power Electronics, 32*(2), 906–910. ISSN: 0885-8993. https://doi.org/10.1109/TPEL.2016.2585658
19. Txapartegi, M., & Liao, J. (2017, March). *Gate driver market and technology trends*. Yole Développment. https://www.i-micronews.com/category-listing/product/gate-driver-market-and-technology-trends-2017.html#description
20. Seidel, A., & Wicht, B. (2018). Integrated gate drivers based on high-voltage energy storing for GaN transistors. *IEEE Journal of Solid-State Circuits*, 1–9. ISSN: 0018-9200. https://doi.org/10.1109/JSSC.2018.2866948
21. Qin, S., Lei, Y., Barth, C., Liu, W., & Pilawa-Podgurski, R. C. N. (2015, September). A high-efficiency high energy density buffer architecture for power pulsation decoupling in grid-

interfaced converters. In *Proceedings of IEEE Energy Conversion Congress and Exposition (ECCE)* (pp. 149–157). https://doi.org/10.1109/ECCE.2015.7309682

22. Afridi, K. K., Chen, M., & Perreault, D. J. (2014). Enhanced bipolar stacked switched capacitor energy buffers. *IEEE Transactions on Industry Applications, 50*(2), 1141–1149. ISSN: 0093-9994. https://doi.org/10.1109/TIA.2013.2274633

23. Chen, M., Afridi, K. K., & Perreault, D. J. (2013). Stacked switched capacitor energy buffer architecture. *IEEE Transactions on Power Electronics, 28*(11), 5183–5195. ISSN: 0885-8993. https://doi.org/10.1109/TPEL.2013.2245682

24. Fichtenbaum, N., Giandalia, M., Sharma, S., & Zhang, J. (2017). Half-bridge GaN power ICs: Performance and application. *IEEE Power Electronics Magazine, 4*(3), 33–40. Navitas driver. ISSN: 2329-9207. https://doi.org/10.1109/MPEL.2017.2719220

25. *LMG3410 600-V 12-A integrated GaN power stage* (2017, April). Texas Instruments Incorporated.

26. Semiconductor, Dialog. (2016). *DA8801 SmartGaN^{TM} integrated 650V GaN half bridge power IC*. Dialog Semiconductor. https://www.dialog-semiconductor.com/sites/default/files/da8801_smartgan_product_brief.pdf

27. Rose, M. Wen, Y., Fernandes, R., Van Otten, R., Bergveld, H. J., & Trescases, O. (2015, May). A GaN HEMT driver IC with programmable slew rate and monolithic negative gate-drive supply and digital current-mode control. In *Proceedings of the IEEE 27th International Symposium on Power Semiconductor Devices IC's (ISPSD)* (pp. 361–364). https://doi.org/10.1109/ISPSD.2015.7123464

28. Moench, S., Costa, M., Barner, A., Kallfass, I., Reiner, R., Weiss, B., et al. (2015, November). Monolithic integrated quasi-normally-off gate driver and 600 V GaN-on-Si HEMT. In *Proceedings of the IEEE 3rd Workshop Wide Bandgap Power Devices and Applications (WiPDA)* (pp. 92–97). https://doi.org/10.1109/WiPDA.2015.7369264

29. *EPC2112 – 200 V, 10 A integrated gate driver eGaN® IC – preliminary datasheet* (2018, March). Efficient Power Conversion Corporation.

30. Moench, S., Reiner, R., Weiss, B., Waltereit, P., Quay, R., Kaden, T., et al. (2018, June). Towards highly-integrated high-voltage multi-MHz GaN-on-Si power ICs and modules. In *Proceedings of the Renewable Energy and Energy Management PCIM Europe 2018; International Exhibition and Conference for Power Electronics, Intelligent Motion* (pp. 1–8).

31. Zhu, M., & Matioli, E. (2018, May). Monolithic integration of GaN-based NMOS digital logic gate circuits with E-mode power GaN MOSHEMTs. In *Proceedings of the IEEE 30th International Symposium on Power Semiconductor Devices and ICs (ISPSD)* (pp. 236–239). https://doi.org/10.1109/ISPSD.2018.8393646

32. Ujita, S., Kinoshita, Y., Umeda, H., Morita, T., Kaibara, K., Tamura, S., et al. (2016, June). A fully integrated GaN-based power IC including gate drivers for high-efficiency DC-DC converters. In *Proceedings of the IEEE Symposium on VLSI Circuits (VLSI-Circuits)* (pp. 1–2). https://doi.org/10.1109/VLSIC.2016.7573496

33. Kaufmann, M., Lueders, M., Kaya, C., & Wicht, B. (2020). 18.2 a monolithic E-mode GaN 15W 400V offline self-supplied hysteretic buck converter with 95.6% efficiency. In *Proceedings of IEEE International Solid-State Circuits Conference – (ISSCC)* (pp. 288–290).

34. Herzer, R. (2010, March). Integrated gate driver circuit solutions. In *2010 6th International conference on Integrated power electronics systems (CIPS)* (pp. 1–10).

35. Zhang, Z. L., Dong, Z., Hu, D. D., Zou, X. W., & Ren, X. (2017). Three-level gate drivers for eGaN HEMTs in resonant converters. *IEEE Transactions on Power Electronics, 32*(7), 5527–5538. ISSN: 0885-8993. https://doi.org/10.1109/TPEL.2016.2606443

36. Seidel, A., Costa, M. S., Joos, J., & Wicht, B. (2015). Area efficient integrated gate drivers based on high-voltage charge storing. *IEEE Journal of Solid-State Circuits, 50*(7), 1550–1559. ISSN: 0018-9200. https://doi.org/10.1109/JSSC.2015.2410797

37. Kaufmann, M., Seidel, A., & Wicht, B. (2020, March). Long, short, monolithic – The gate loop challenge for GaN drivers: Invited paper. In *Proceedings of IEEE Custom Integrated Circuits Conference (CICC)* (pp. 1–5). https://doi.org/10.1109/CICC48029.2020.9075937

Chapter 2
Fundamentals

2.1 Gate Drivers and Power Stages

2.1.1 Driver Configurations and Building Blocks

Figure 2.1 shows fundamental power switch and gate driver configurations [1]. The power switches symbols represent various kinds of power transistors, like metal oxide semiconductors (MOS), gallium nitride (GaN), or insulated gate bipolar transistors (IGBT). The power transistor along with the gate driver can be implemented at the low-side (Fig. 2.1a) or at the high-side (Fig. 2.1b), respectively. Both switches together form a half-bridge (Fig. 2.1c) and two half-bridges a full-bridge, also called H-bridge. Loads can be, for example, motors, transformers, filter inductors, resonance inductors/capacitors of a DCDC converter.

A complete driver design includes several circuit blocks; some are highlighted in Fig. 2.1. The control logic delivers the turn-on or turn-off signal for the power stage, usually in a pulse width modulated fashion. A level shifter converts the driver control signal from the low-voltage domain (referred to the system ground) into the driver voltage domain (referred to the source of the power transistor). The level shifter and driver supply may include a galvanic isolation for reasons, described in Sect. 2.4.1. Depending on application and requirements, also the low-side gate driver may need a galvanic isolated level shifter and power supply [2].

2.1.2 Gate Driver Output Stage

Integrated gate drivers in power applications in the kW-range drive large power switches. Fast switching and a high integration level is key for highly efficient, compact, and reliable power electronics. Gate voltages in the range of $-8\,\text{V}$ to $\sim15\,\text{V}$ for Si or SiC and $-5\,\text{V}$ to $6\,\text{V}$ for GaN transistors demand for typically

© The Author(s), under exclusive license to Springer Nature Switzerland AG 2021
A. Seidel, B. Wicht, *Highly Integrated Gate Drivers for Si and GaN Power Transistors*, https://doi.org/10.1007/978-3-030-68940-7_2

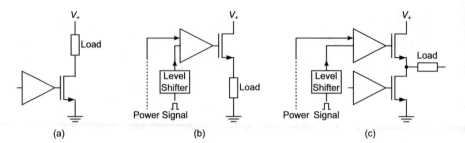

Fig. 2.1 Driver configuration in application as (**a**) low-side driver, (**b**) high-side driver, and (**c**) low- and high-side driver in a half-bridge

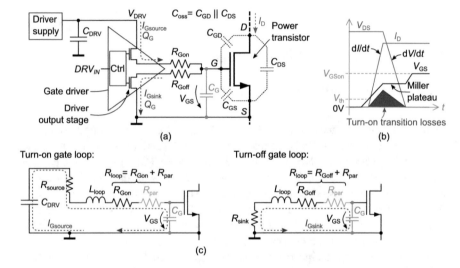

Fig. 2.2 (**a**) Conceptual circuit of a generic gate driver, (**b**) power transistor switching transition, and (**c**) turn-on and turn-off gate loop

~30 V-maximum-rated gate driver output stages. For a unipolar gate control, a maximum-rated driver output voltage of 15 V may be sufficient. Figure 2.2 shows a typical setup of a gate driver and power switch. In fully integrated gate drivers, large die area is occupied by the driver output stage consisting of two or more high-voltage transistors (rated for 15 V and above). Figure 2.3 shows possible configurations for driver output stages using CMOS devices [3]. The CMOS inverter in Fig. 2.3a is a simple and often used configuration. Since a PMOS transistor is typically about three times larger than an NMOS transistor with the same on-state resistance, the output stage gets large, especially for large output currents. Furthermore, many technologies do not offer area-efficient high-voltage PMOS devices [4]. On IC-level, transistors with voltage ratings of 20 V count as high-voltage devices. In some high-voltage technologies, the high-voltage PMOS transistor can be up to 30 times larger in its specific resistance compared to the

Fig. 2.3 Three configurations for output stage buffers with (**a**) pull-down NMOS transistor and pull-up PMOS transistor, (**b**) two NMOS transistors, and (**c**) NMOS pull-down and NMOS/PMOS pull-up transistors

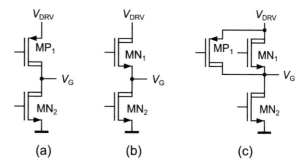

NMOS transistor. Therefore, Fig. 2.3a is more suitable for output stages with smaller output currents [5].

To save area and cost, an output stage with two NMOS transistors can be used [4, 6, 7], Fig. 2.3b. A bootstrap circuit is required to provide the gate overdrive voltage for the NMOS transistor in the pull-up path (MN_1). Usually, the bootstrap capacitor C_{B1} is large [8]. Therefore, the bootstrap capacitor is often implemented as a discrete component because of its size [6]. The function of the two NMOS output stage is explained further below in this section.

Figure 2.3c shows a way to avoid a bootstrap circuit with a PMOS transistor in parallel to the NMOS pull-up transistor [9, 10]. During turn-on, the gate node of MN_1 is connected to V_{DRV}, carrying the large drive current in the beginning of the switching phase. When the output node V_G approaches V_{DRV}, the gate-source voltage decreases and finally turns-off MN_1. MP_1 has to pull-up and clamp the output node fully to V_{DRV} and can be smaller than in the output stage configuration of Fig. 2.3a. Nevertheless, it should be large enough to maintain fast switching, especially at small driver voltages V_{DRV}. To keep the driver strength even for lower V_{DRV}, a modified version of the output stage of Fig. 2.3c is presented in [10]. The circuit is optimized for speed, but still requires high-voltage PMOS transistors, a smaller but not negligible bootstrap capacitor and a complex control circuit.

The choice of the most suitable output stage configuration strongly depends on technology, especially regarding the size of high-voltage PMOS transistors and the capacitance density of capacitors. To save area and cost, an output stage with two NMOS transistors based on Fig. 2.3b is utilized in Chap. 3.

Figure 2.4 shows the basic implementation of a conventional gate driver based on an output stage with two n-channel transistors (see Fig. 2.3b). A conventional bootstrap circuit provides the gate overdrive for turning on the high-side transistor MN_1. If $DRV_{IN} = 0\,\text{V}$, the gate driver is in off-state, and the node V_G is shorted to ground by MN_2. V_1 charges C_{B1} to 5 V minus the forward voltage of D_{B1}. By setting DRV_{IN} to high, MN_2 turns off and MN_1 is switched on by the level shifter signal $LSout$ and the pre-driver (buffer) B_2. The V_G node rises to V_{DRV}, while D_{B1} prevents C_{B1} from discharging to V_1. V_{B1} serves as floating voltage supply and has to provide the charge Q_{tot} for B_2, the gate capacitance of MN_1, and the level shifter. A voltage dip occurs at C_{B1} after the discharge process to deliver the

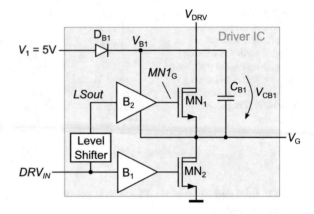

Fig. 2.4 Two-NMOS transistor output stage buffer with a conventional bootstrap circuit

gate charge for MN_1 and the high-side circuits, like the level shifter and the buffer B_2. The charge that is available from C_{B1} to achieve a voltage dip ΔV_1 at V_{B1} can be calculated as $Q_{tot} = C_{B1} \cdot \Delta V_1$. For MOSFET buffers such as the buffer in Fig. 2.4, a typical undervoltage level is in the $4 \sim 5$ V range, assuming a nominal gate-source voltage of 5 V for MN_1, MN_2. An upper voltage limit is given by the breakdown voltage of the V_{B1} rail, which is typically $\sim$5.5 V for such transistors. For circuits with an integrated C_{B1}, often undervoltage levels down to $\sim$3.5 V are chosen [10] to decrease the size of C_{B1}. This is advantageous, even with an enlarged MN_1 to achieve the same MN_1 driving capability. Nevertheless, a bootstrap capacitor charged to $\sim$5 V can only be discharged by about $\Delta V_1 = 0.5$ V...1.5 V. The conventional bootstrap circuit suffers from small charge allocation with respect to the whole stored charge in the bootstrap capacitor. This is addressed in Chap. 3, which presents a gate driver with area-efficient implementation of a output stage with two NMOS.

2.1.3 Basic Gate Driver Operation

Figure 2.2a shows the circuit principle of a gate driver. To turn-on the power transistor, its gate gets connected to V_{DRV}, while its gate is pulled to the source potential S to turn it off. The gate of the power transistor represents a capacitive load. The larger the gate currents $I_{G,source}$ and $I_{G,sink}$, the faster the rate of change of the gate voltage V_{GS}. The gate currents are limited by any parasitic resistance in the turn-on/-off gate loop, like the on-state resistances of the gate driver output stage transistors or the internal gate resistance of the power transistor. The value of the internal gate resistance can be taken from the data sheet of the power transistor like in [11]. In case of discrete power transistors, any bond wires and other package interconnects in the gate loop are further limiting the transient gate current.

The gate resistors $R_{G,on}$ and $R_{G,off}$ are implemented to adapt the gate currents and thus the gate drive speed. This also prevents ringing or a gate voltage overshoot at V_{GS} [12–14]. On-chip power transistors are, typically, driven without additional gate resistors, while the gate current is only determined by the on-state resistances of the last gate driver stage. Section 2.1.4 gives more details about the impact of gate resistors.

Figure 2.2b shows the basic transient turn-on behavior of the power transistor. When the gate voltage V_{GS} crosses the threshold voltage V_{th}, the drain current I_D rises, until V_{GS} reaches the Miller plateau level. During the Miller plateau phase, the transistor discharges its output capacitance C_{oss} (parallel connection of C_{DS}, C_{GD}) and all other capacitances connected at the switching node depending on application. During that phase, the gate driver provides the current for discharging C_{GD}, without charging C_{GS}, leading to a constant V_{GS}, until V_{DS} is at its minimum value. Hence, the strength of the gate driver essentially determines the transition speed dI/dt of I_D and the dV/dt of V_{DS}. A description of the losses during the switching transition is given in Sect. 2.1.8.

2.1.4 Gate Loop Parasitics

Figure 2.2c shows the gate loop that forms a (parasitic) resonance circuit at driver turn-on and turn-off, respectively. The resonance circuit consists of the parasitic gate loop inductance L_{loop} mainly caused by interconnections between the components in the gate loop, the gate capacitance C_G that is in series with C_{DRV} in case of a turn-on loop, and R_{loop}. R_{loop} comprises the resistance of the gate resistors $R_{G,on}$ and $R_{G,off}$, respectively, and any other parasitic resistances R_{par} in the gate loop. R_{par} consists, for example, the channel resistance of the driver output stage transistors and the internal gate resistance of the power transistor.

To prevent the gate voltage V_{GS} from oscillating, or overshooting and undershooting, the gate loop must be damped. This is typically done by the gate resistors $R_{G,on}$ and $R_{G,off}$ that contribute to the overall gate resistance R_{loop} [12–14]. R_{loop} dissipates the energy stored in the gate loop inductance L_{loop} at the end of a gate charge or discharge process. The damping factor D and the quality factor Q, respectively, are measures for the damping of the gate loop [14]:

$$D = \frac{R_{loop}}{2} \cdot \sqrt{\frac{C_{loop}}{L_{loop}}} = \frac{1}{2 \cdot Q} \tag{2.1}$$

$D = 1$ critically damps the gate loop, i.e., no V_{GS} overshoot occurs. In this case, the minimum allowed gate resistance is

$$R_{loop} = 2 \cdot D \sqrt{\frac{L_{loop}}{C_{loop}}}. \tag{2.2}$$

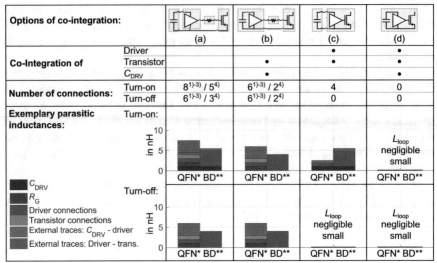

Options of co-integration:		(a)	(b)	(c)	(d)
Co-Integration of	Driver			•	•
	Transistor		•	•	•
	C_{DRV}		•		•
Number of connections:	Turn-on	$8^{1)-3)}$ / $5^{4)}$	$6^{1)-3)}$ / $2^{4)}$	4	0
	Turn-off	$6^{1)-3)}$ / $3^{4)}$	$6^{1)-3)}$ / $2^{4)}$	0	0

Exemplary parasitic inductances:	Turn-on:				

Turn-on chart (in nH, scale 0–10):
QFN* BD** (a) | QFN* BD** (b) | QFN* BD** (c) | QFN* BD** (d): L_{loop} negligible small

Legend:
- C_{DRV}
- R_G
- Driver connections
- Transistor connections
- External traces: C_{DRV} - driver
- External traces: Driver - trans.

Turn-off: (in nH, scale 0–10):
QFN* BD** (a) | QFN* BD** (b) | QFN* BD** (c): L_{loop} negligible small | QFN* BD** (d): L_{loop} negligible small

* QFN driver package, ** Bare die driver & transistor, without R_G (driver & transistor directly bonded)
See Appendix 2.A for legend of labels "1)-4)" and for more details

Fig. 2.5 Exemplary gate loop inductance values and number of interconnections assuming gate drive integration options (a) to (d) of buffer capacitor, gate driver, and power transistor

Choosing $D = 0.7$ ($Q \sim 0.7$) leads to a slight V_{GS} overshoot at turn-on or undershoot at turn-off. Higher values for $R_{G,on}$ and $R_{G,off}$ are required in case of a large L_{loop}, slowing down the gate charge/discharge speed, which will be discussed in more detail in Sect. 4.1.2. To keep the inductance small, effort has to be spent on assembly and layout design [13, 15, 16]. This is the case especially on substrates with only a single wiring layer, which is typical for layouts on ceramic substrates. Chapter 5 presents a fast switching gate driver approach that can handle large gate loops.

Figure 2.5 shows various integration options of the gate driver: (a) buffer capacitor C_{DRV}, gate driver, and power transistor on separate dies; (b) C_{DRV} and gate driver on the same die; (c) gate driver and power transistor on the same die; and (d) C_{DRV}, gate driver, and power transistor on the same die. Figure 2.5 considers typical gate loop inductances for the gate driver and power transistor in a quad-flat no-lead package (QFN) package and as bare dies, which are often used package options. Appendix 2.4.5 shows some more package scenarios. Main parts of the total gate loop inductance L_{loop} are shown based on the exemplary values derived in Appendix 2.4.5. The number of connections between the components for each option is an indicator for the reliability of a system [17].

Integration Option Fig. 2.5a All components are implemented as separate integrated circuits (ICs), which is currently the most widespread option providing high flexibility. However, the amount of interconnections results in relatively large parasitic gate loop inductances and in slower gate drive speed, as mentioned earlier in this section. Chapter 5 presents a gate driver architecture based on the concept of high-voltage energy storing (HVES) that achieves fast and robust switching even at large gate loop inductances.

Integration Option Fig. 2.5b This option integrates the driver buffer capacitor C_{DRV} on-chip. A C_{DRV} integration leads to fewer connections and off-chip components increasing the overall reliability. The decrease of the turn-on gate loop inductance in Fig. 2.5c is only marginal. However, it considers an optimized placement of C_{DRV} next to the gate driver, which is not possible in all cases as mentioned above. A more compact, more flexible, and more easier gate driver assembly is possible with the absence of an external buffer capacitor. This work contributes to the full integration of C_{DRV} by area-efficient buffer integration methods, called high-voltage charge storing (HVCS) (Chap. 3) and HVES (Chap. 4).

Integration Option Fig. 2.5c A monolithic co-integration of the power transistor and the gate driver significantly reduces the parasitic inductances. Due to the fully integrated turn-off gate loop, no parasitic inductance is caused by interconnections anymore. The on-chip wires have negligible influence as they are very small. This is one reason why a co-integration is very attractive in line with the common trend for GaN power transistors [18–28].

Advantages of a co-integration of the power transistor and gate driver:

- Robust switching, because of negligible parasitic gate loop inductance in the turn-off gate loop (see Sect. 2.3);
- Easy to implement as it can be controlled by logic levels;
- Optimized gate driver design for power switch;
- No bulky gate driver IC next to the power transistor

Disadvantages of a co-integration of the gate driver and the power transistor:

- Limited modularity: The gate driver cannot be chosen and typically not be parametrized according to application requirements. Special functions may not be available, like thermal shutdown, active gate control (see Sect. 2.3.2), current sensing with over-load detection, etc.
- Less flexible controllability and optimization possibilities: The gate drive speed cannot be easily changed via a gate resistor. Furthermore, the process technology is typically optimized for the power transistor and not for the gate driver [29]. This limits the performance and functionality of the gate driver. For example, gate drivers on a GaN substrate are currently limited to very simple circuit structures [25–27] with disadvantages like a high static power consumption, a slow gate charge process, or no rail-to-rail gate drive output.
- As many silicon power transistors are vertical devices, i.e., their source are at the wafer's top and the drain at their bottom side [30] (see Sect. 2.2.1), effort has to be spent for adding a gate driver circuit in CMOS [17]. Thus, the monolithic co-integration of the power transistor and gate driver is typically better suited for lateral devices [28, 31, 32], which is the case for silicon transistors with lower power ratings, and GaN transistors.

Integration Option Fig. 2.5d The integration of all components on the same die avoids any off-chip connections in the gate loop [28, 32]. The parasitic inductances are negligible, achieving robust switching. Due to the low parasitic inductance also of the turn-on gate loop, it is the best case for fast switching. The area-efficient capacitor integration concept of HVCS in Chap. 3 presents a solution that requires no inductance in the gate loop, which is very attractive supporting the trend towards monolithic integration. The concept of HVES, utilizing a small stacked on-chip inductor (Chap. 4), is very suitable, as well. It achieves an area-efficient capacitor integration with all the advantages of fast switching characteristic of the HVES concept (see Sect. 4.1.2) and the absence of an external component.

2.1.5 Buffer Capacitor CDRV

The gate driver supply (Fig. 2.2) needs to be buffered with a bypass capacitor (buffer capacitor), which is placed near to the gate driver IC, to reduce the turn-on gate loop inductance. In case of a bootstrap driver supply (see Sect. 2.4.2), the buffer capacitor provides the gate charge during gate driver on-state when the driver supply is not available. For fast gate charge events, it can be assumed that C_{DRV} delivers all the gate charge during driver turn-on, charging the gate capacitance C_G of the power device. This causes a voltage drop ΔV_C at C_{DRV}. Typically, $\Delta V_C \ll 0.5$ V is required to maintain an appropriate gate driver supply. The gate charge expression

$$Q_G = C_{DRV} \cdot \Delta V_C \qquad (2.3)$$

indicates that, with on-chip capacitors, only a small amount of charge can be delivered. For example, the gate charge of a GaN power transistor with $Q_G = 5$ nC should be delivered. $\Delta V_C = 0.5$ V would require $C_{DRV} = 10$ nF, a typical capacitance density of 0.4 nF mm^{-2} of a 5 V-rated poly-nwell capacitor, on a die area of 24 mm^2. Trench capacitors offer a significantly higher capacitance density. However, in many technologies they are not available or cause additional costs. Therefore, C_{DRV} is typically placed off-chip. Chapters 3 and 4 propose area-efficient methods for integrating buffer capacitors on-chip, called high-voltage charge storing (HVCS) [3] and high-voltage energy storing (HVES) [33, 34]. The gate driver based on the concept of HVCS is developed for silicon transistors, which have a gate charge, which is ~10x higher than for GaN counterparts. Therefore, in Chap. 3, the HVCS capacitor implementation method is applied to the internal pre-driver, driving the 0.3 nC high-side transistor of the gate driver output stage. Gate drivers based on HVES (see Chap. 4) are capable for higher gate charge delivery, demonstrated up to 11.6 nC, which enables to integrate the buffer capacitor of a gate driver for most commercially available GaN transistors.

2.1.6 Gate Driver Types

A voltage-source controlled gate driver (Fig. 2.2a), also referred to as hard-switching gate driver [1], is the most commonly used gate driver type. Typically, the gate current is determined by the gate resistor. In contrast, gate drivers with current-source behavior have no gate resistors. The gate current is controlled in first approximation independently from any parasitics in the gate loop [35]. This can be achieved by operating the driver output stage transistors in saturation, i.e., as a current source [36–38]. Alternatively, there are resonant gate drive concepts that can be considered to be current source gate drivers, like [9, 39–46]. This is because they charge the gate via a resonance inductor that has current source behavior. The HVCS gate driver in Chap. 3 can be categorized as a charge-source controlled gate driver. It can be seen as a subcategory of the current-source controlled gate drivers, since the output stage transistor is predominantly operated in saturation. The HVES gate drivers in Chaps. 4 and 5 provide a certain amount of energy to charge the gate. This type of gate driver can be described as energy-source controlled gate driver, which is another subcategory of the current-source controlled gate drivers, because of its resonant behavior with an inductor in the gate charge path.

2.1.7 Efficiency of Resonant and Non-resonant Gate Drivers

According to Sects. 2.1.3 and 2.1.6, in non-resonant gate drivers the resistive components dominate in the gate charge path, while the inductive part L_{loop} should be as small as possible. A gate charge of $Q_G = C_G \cdot V_{DRV}$ flows into the gate. Assuming that Q_G is drawn from V_{DRV} results in an energy delivery from V_{DRV} of

$$E_{in} = Q_G \cdot V_{DRV}. \tag{2.4}$$

The well-known expression for a first-order approximation of the stored energy in the charged gate capacitance C_G is

$$E_{out} = 0.5 \cdot C_G \cdot V_{DRV}^2, \tag{2.5}$$

assuming a voltage-independent C_G. Figure 2.6a shows the charge-voltage (QV) diagram, charging C_G from zero to V_{DRV} with a negligible L_{loop} in the gate loop. Q_G increases proportionally to the gate voltage across the capacitance C_G. The triangular area above the Q_G-curve represents the stored energy E_{out} in C_G. At the beginning of the charging event ($V_{GS} = 0\,V$), V_{DRV} fully drops across the gate loop resistance R_{loop}, which corresponds to large power loss in R_{loop}. With further increasing V_{GS}, the voltage drop across R_{loop} decreases and lower power losses occur. Integrating the total power loss in R_{loop} during the charging event results in the dissipated energy in R_{loop}, which can be interpreted as the area below the

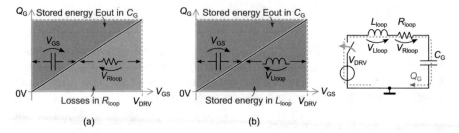

Fig. 2.6 QV-diagrams and the equivalent circuit for charging C_G via a (**a**) non-resonant gate driver and (**b**) resonant gate driver

Q_G-curve in Fig. 2.6a. This leads to energy losses of a gate driver turn-on event:

$$E_{\text{loss}} = E_{\text{in}} - E_{\text{out}} = 0.5 \cdot C_G \cdot V_{\text{DRV}}^2 \tag{2.6}$$

Thus, always half of the energy is dissipated, resulting in a turn-on efficiency of a non-resonant gate driver of 50 %. This is independent of the resistance R_{loop}. R_{loop} only determines how fast the energy is dissipated.

At driver turn-off, the energy E_{out} stored in the gate capacitance is dissipated in the resistances of the turn-off gate loop. Hence, the total energy consumption of a non-resonant gate driver can be calculated according to (2.4).

Resonant gate driver concepts, instead, charge the gate via an inductor. For efficiency considerations, it can be added to the overall gate loop inductance L_{loop} and have negligible R_{loop} in the resonance circuit. The corresponding QV-diagram is shown in Fig. 2.6b. During the whole gate charge process, V_{DRV} drops over energy saving components, i.e., over C_G and L_{loop}, that store the energy. At the end of the gate charge process, L_{loop} contains the cumulated energy in form of a magnetic field, keeping the inductor current flowing. Resonant gate drivers recover this energy by commutating the current back into its gate driver supply. In theory, resonant gate drivers can achieve an efficiency of 100 %. Moreover, capacitor charging losses can be avoided by charging it from a current source, for example, via an inductive link with current source characteristic at its output [47]. The current source varies its output voltage, tracking the capacitor voltage V_{GS}, assuming a negligible R_{loop}, with no voltage drop across it.

In real world, the charging event is neither purely resonant nor purely non-resonant. Non-resonant gate drivers contain always some parasitic gate loop inductance L_{par} in the gate loop. However, since this energy is not recovered, the energy is dissipated in R_{loop}. In case of a relatively fast gate charge process, the remaining energy in L_{par} overcharges the gate and flows back again, until all energy is dissipated in R_{loop}, leading to the well-known effects of a gate overshoot or even ringing. Resonant gate drivers have as well an R_{loop} in the gate loop, which decreases the gate drive efficiency. Typically, larger inductances in the gate loop improve their quality factor (see Sect. 2.1.4), improving the efficiency of resonant gate drivers, but decrease the gate drive speed [46].

2.1.8 Power-Transistor Switching Losses

During a typical switching transition, as shown in Fig. 2.2b, various kinds of losses occur every switching cycle and increase linearly with the switching frequency of the power transistor. These losses are called dynamic or switching losses. Lower switching frequencies decrease the dynamic losses, which allows for better application efficiencies and miniaturization due to less cooling effort. Or, higher switching frequencies in the application can be applied, scaling down the passives. A trade-off has to be found. Low dynamic losses are a key factor for compact power electronics applications.

The following losses can be distinguished.

Transition Losses For the short time of transition, the multiplication of I_D and V_{DS} results in a short high power loss peak (power loss triangle), as depicted in Fig. 2.2b. The area under the power loss triangle is the dissipated energy. The amount of these losses can be different for a power transistor turn-on and turn-off events. In case of a fast turn-off, these losses are typically much lower or can be almost eliminated, since the gate driver turns-off the power transistor, before V_{DS} starts to rise. Since these losses contribute significantly to the overall application efficiency, fast turn-on and turn-off transitions, and thus fast gate drivers, are crucial to minimize the area under the loss triangle. To comply with electromagnetic compatibility (EMC) requirements, often a trade-off has to be found between losses and switching speed reduction [36, 37]. In soft-switching applications, the transition losses can be eliminated, which is discussed further below.

Parasitic Transistor Capacitances Losses occur by charging and discharging the parasitic transistor capacitances. As described in Sect. 2.1.7, the gate capacitance leads to losses in the gate driver. The output capacitance C_{oss} ($C_{GD} + C_{DS}$) causes losses as well, if it is charged or discharged over a current path, which is mainly determined by loss dissipating elements, like a resistor, a transistor channel, parasitic resistances, or a diode (see Sect. 2.1.7). For example, in a hard-switching half-bridge application, one transistor is in the off-state, while its C_{oss} is charged via the transistor channel of the other half-bridge transistor. During a transistor turn-on event, the transistor shortens C_{oss} and dissipates the stored energy in the transistor channel. Losses caused by C_{oss} can be minimized in soft-switching applications, which is described below. For loss calculations, it is crucial to consider the typically high voltage dependency of C_{oss} (see Appendix 2.4.5).

Reverse Conduction Losses Depending on the application, every switching cycle the power transistor is in reverse conduction, i.e., for silicon transistor its body diode or for GaN transistors its channel (quasi-body diode). The quasi-body diode behavior is further explained in Sect. 2.2.3. The current through this diode or channel leads to losses, due to the forward voltage drop across it.

Reverse Recovery Losses For silicon transistors, after the reverse conduction phase, the body diode should block the current, but in reality it still conducts for a certain time. For a short time, cross currents can occur in an application, which are responsible for further significant losses.

The typical turn-on transition, shown in Fig. 2.2b, is called hard-switching, since the current turns on, while V_{DS} is still at its maximum value (transistor turn-on event) or the current is still flowing, while V_{DS} starts to rise (transistor turn-off event). This causes many losses, as explained above.

Many applications use soft-switching techniques to minimize these losses. As shown in Fig. 2.7, soft-switching (zero-voltage switching) is a technique to resonantly discharge C_{oss} before turning-on the power transistor. For turn-off, it is vice versa, i.e., the power transistor is turned-off before C_{oss} is charged. The transistor switches always at $V_{DS} = 0\,V$, eliminating the transition losses. The resonant charging/discharging of C_{oss}, e.g., via an inductive load, minimizes the losses caused by this parasitic capacitance. Some losses remain due to the non-ideal resonance, similar to the gate driver losses in a resonant gate driver (see Sect. 2.1.7). For considerations to soft-switching in a GaN transistor half-bridge configuration, see Sect. 2.3.2. However, soft-switching cannot always be applied or not fully applied to an application and, because of other disadvantages, is not suitable in any cases.

2.2 Power Transistors and Applications

2.2.1 Silicon Transistors

Silicon is a well-known material for power transistors. Many years of application, even in harsh environments, give detailed experiences regarding reliability, device and product testing conditions, and circuit development. Long terms of research and manufacturing lead to highly performing and cost-effective devices. Figure 2.8 shows some important types of silicon transistors that are typical for applications in the lower kW-range.

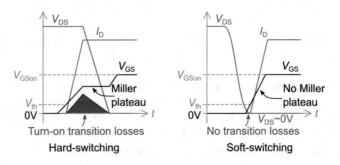

Fig. 2.7 Transitions of a hard-switching versus soft-switching power transistor

Planar silicon Power MOSFET The cross section of a planar silicon power transistor is shown in Fig. 2.8a. It is a vertical device with the source at the top and the drain at the bottom side of the wafer. A gate voltage above the device threshold voltage generates a conducting n-type inversion channel under the gate in the p^+-doped region turning the transistor on. An increase of the breakdown voltage during off-state can be achieved by low-doping the n_{epi} drift region and a high thickness of this region. However, this increases the device on-resistance. The theoretically achievable specific on-resistance of a silicon power MOSFET is limited for a given voltage class [30, 48]. Silicon power transistors are typically available with voltage ratings up to 200 V.

Superjunction MOSFET (SJMOSFET) Figure 2.8b shows a silicon SJMOSFET which overcomes the limitation of the planar silicon power transistor. The superjunction technology is based on two key principles. First, the n_{epi} region is much more heavily doped compared to a conventional silicon MOSFET power transistor leading to a lower specific on-resistance. However, to still obtain a high blocking voltage capability, p-doped trenches are added that provide a compensating space charge region at its junction to the n_{epi} [48]. SJMOSFETs operate mainly in the range of 200 V − 900 V. However, the borders move towards higher and lower voltages [50].

IGBT Figure 2.8c shows a cross section of an IGBT. Although the IGBT differs from a silicon power MOSFET only by an additional p_+ layer beneath the n_{epi} region, it behaves more like a bipolar transistor. The IGBT is a bipolar device because the p_+ collector injects minority carriers into the n-region during on-state. Silicon power MOSFETs and SJMOSFETs are unipolar devices. The equivalent device circuit structure is a combination of a pnp bipolar transistor controlled via an n-channel MOSFET leading to an insulated gate structure. The IGBT supports very high currents and low on-resistances. However, as it is a minority carrier device, the switching performance is below that of a silicon MOSFET [51]. IGBTs are attractive for the low-medium voltage range of 400 V − 1700 V for many applications like for the consumer market because they are cheaper than SJMOSFETs and show

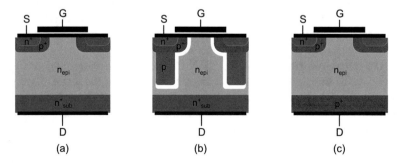

Fig. 2.8 Basic cross sections of (**a**) a standard power MOSFET, (**b**) SJMOSFET, and (**c**) IGBT, based on [48, 49]

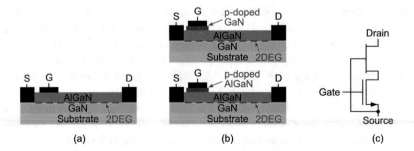

Fig. 2.9 Cross section of (**a**) basic depletion-mode GaN transistor, (**b**) two types of enhancement-mode GaN transistor, and (**c**) a circuit of a cascode device

sufficient performance [51, 52]. More information about the function and the characteristics of IGBTs is given in [49].

2.2.2 GaN Transistor

In recent years, wide-bandgap materials, in particular gallium nitride (GaN) and silicon carbide (SiC), came into focus for power transistors. GaN offers significantly better electrical characteristics than silicon and SiC [52] and is on the way to become the next-generation power electronics material [53]. The advantage of SiC lies in its higher thermal conductivity [52, 54]. As both GaN and SiC have a wider band gap, compared to silicon [55, p. 25], higher energy is necessary to raise a valence electron into the conduction band. This allows to operate the transistor at higher junction temperatures. Furthermore, a wider band gap allows higher critical electric field strength leading to a higher breakdown voltage at a certain channel length of the semiconductor, which means a better achievable specific on-resistance. As the electron mobility is twice as high, GaN offers a smaller specific on-resistance compared to silicon [55, p. 41]. The GaN transistor channel between drain and source is commonly based on an AlGaN/GaN heterostructure (see this section below). During GaN transistor on-state, a highly-conductive two-dimensional electron gas (2DEG) is formed in this region further increasing the electron mobility compared to a pure GaN material [56].

Figure 2.9 shows various types of GaN transistors. Currently, they are lateral devices, i.e., its drain and source are both on the top side of the wafer. Figure 2.9a shows a basic and simplified GaN transistor cross section, while real devices have several more layers that are important for manufacturing, reliability, and good device characteristics. A characteristic highly-conductive two-dimensional electron gas (2DEG) is formed in the AlGaN/GaN drift region [31]. Hence, the GaN transistor is naturally a normally-on (depletion-mode) device. For turning off the device, a simple metal layer as gate on top of the AlGaN layer requires a negative V_{GS} (typically by a few volts) to deplete the 2DEG. The normally-on behavior is an

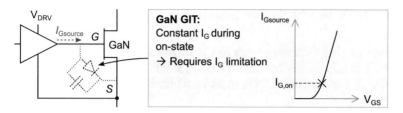

Fig. 2.10 GaN GIT gate characteristics

issue for most applications since they need blocking devices when the gate control is not available, e.g., during application start-up or failure. Figure 2.9b shows two popular solutions to get a normally-off (enhancement-mode) transistor by inserting a p-doped layer between the gate and the AlGaN layer [31, 57–59]. The p-doped layer depletes the 2DEG at a device threshold voltage of $V_{GS} \sim 1$ V. EPC first commercialized the eGaN®-FET with a p-doped GaN layer. Panasonic sales the gate injection transistor (GIT) with p-doped AlGaN layer beneath the gate. Both transistor types have gates with a current characteristic similar to a diode (Fig. 2.10). During transistor on-state, the gate is slightly in forward conduction to maintain the low-ohmic GaN on-state resistance [57]. The gate leakage can be up to ~30 mA depending on the device size, but it is generally higher for GaN GIT transistors [60, 61]. Therefore, GaN transistors with a diode behavior and significant gate leakage are referred as gate-injection transistors (GIT) in this work, representing devices of various manufacturers. The injected holes into the gate reduce the $R_{DS,on}$ increasing the device performance [57, 59]. However, there are currently also devices with an isolated gate structure available with significantly lower gate leakage [11, 57, 58, 62].

Figure 2.9c shows a cascode comprising a high-voltage blocking normally-on GaN transistor, which gets normally-off behavior in conjunction with a low-voltage silicon transistor. If the silicon transistor is not controlled, the gate of the GaN device is pulled negative turning off the GaN. When the silicon device is in on-state, it shortens the gate-source terminals of the GaN transistor turning it on. Normally-on GaN devices have performance and manufacturing advantages compared to normally-off counterparts [63]. They show better performance, because they start with full charge under the gate, which leads to high conductivity within a certain die area. Current normally-off devices are designed with less charge under the gate to achieve device off-state at a gate voltage of 0 V [64]. However, the cascode in this form is not well controllable as the gate of the GaN is typically not accessible and it tends to have higher parasitic inductances due to more interconnections [57]. Brohlin et al. [65] and Norling et al. [66] present cascodes where the gate of the normally-on device is directly driven and the silicon transistor, kept in on-state during normal operation, is included only for safety reasons. Norling et al. [66] is about normally-on SiC transistors with similar issues. Cascodes are more effective for GaN transistors with higher blocking voltages than the silicon transistor because the impact of the silicon device on the overall $R_{DS,on}$ gets lower [67].

2.2.3 GaN Versus Silicon Transistors

This section shows characteristics of GaN transistors and the application impact compared to silicon devices, mainly SJMOSFETs. This work focuses mainly on GaN transistors and SJMOSFETs as they are both fast switching devices in contrast to IGBTs.

2.2.3.1 Losses

Minimum losses are crucial for both high power conversion efficiencies and highly compact power electronics (see also Sect. 2.1.8). There are essential differences between the losses that are caused by a GaN transistor in contrast to a silicon transistor. This will be discussed below.

Parasitic Capacitances The 2DEG in the GaN transistor leads to a highly conducting channel that contains a very low amount of charge. As shown in Fig. 2.11, this leads to a very low specific on-resistance and very low parasitic capacitances compared to silicon devices indicated by the gate charge of Q_G. Although generations of silicon transistors became better over the years (see figure of merits (FOMs) of SJMOSFETs of the years 2002 and 2013 in Fig. 2.11), GaN transistors have a naturally better switching FOM. It is 8–25 times higher than for SJMOSFETs and SiC Mosfets [68]. Significant losses in the transistor power stage are caused by its output capacitance C_{oss}, especially in hard switching applications (see Sect. 2.1.8). Although C_{oss} is even higher for GaN transistors compared to modern SJMOSFETs, the losses can be significantly lower, because SJMOSFETs have typically a highly non-linear C_{oss}, which is described in more detail in Appendix 2.4.5. Due to lower gate capacitance and lower gate voltages of about 5 V, GaN transistors require $\sim 10 \times$ lower gate charge Q_G, which drastically lowers the gate driver power consumption. This reduces the size of the driver supply circuits that occupy significant circuit board area.

Switching Transitions The small parasitic capacitances of the GaN transistor with respect to its $R_{DS,on}$ result in very fast switching transitions of up to $1000 \, \text{V ns}^{-1}$ [69–71] and $> 10 \text{A ns}^{-1}$ [72]. While faster switching transitions reduce the transition losses (see Sect. 2.1.8), they bring challenges in terms of robust switching [12]. Modern SJMOSFETs have an even lower C_{oss} compared to GaN transistors, as shown in Fig. 2.11, resulting in very short transition times as well. However, silicon MOSFETs have a parasitic bipolar transistor that limits their transition speed [73]. The limit is usually reported in the datasheets. The highly non-linear C_{oss} of SJMOSFETs results in a varying dv/dt of the voltage transition making the timing and dv/dt control more challenging [74]. The switching slopes of GaN and SJMOSFETs are directly controllable by the gate signal in contrast to IGBTs which have a so called tail current at turn-off [49].

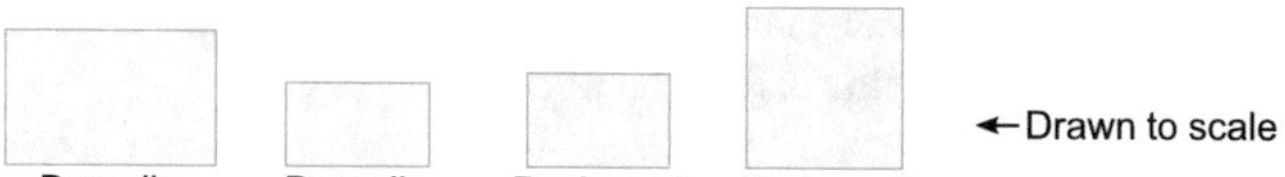

← Drawn to scale

	Bare die Si CoolMOS[1] 2002 (650V, 47A)	Bare die Si CoolMOS[2] 2013 (700V, 33A)	Packaged GaN[3] * (650V, 30A)	Packaged GaN GIT[4] * (600V, 26A)
C_{oss} ($C_{GD}+C_{DS}$):	2200pF	48pF	65pF	71pF
Q_G:	252nC	64nC	6.5nC	5nC
R_{DSon}:	70mΩ	65mΩ	50mΩ	56mΩ
FOM ($Q_G \cdot R_{DSon}$):	18ΩnC	4ΩnC	0.32ΩnC	0.28ΩnC
V_{GS} typ.	10V	10V	5-6V	3.5V
V_{GS} static ratings:	±20V	±20V	-10V...+7V	-
V_{GS} transient ratings:	±30V	±30V	-20V...+10V	-10V
V_{th} (typ / min):	3V / 2.1V	3.5V / 3V	1.7V / 1.1V	1.2V / 0.9V

1) SPW47N60C3
2) IPP65R065C7
3) GS66508T
4) PGA26E07BA
* Unknown year of first generation

Fig. 2.11 Comparison of various silicon SJMOSFETs and GaN transistors

Reverse Conduction Behavior Silicon MOSFETs and SJMOSFETs have a pn-junction between the source and drain (see Fig. 2.8a) forming a so called body diode, which can be used as freewheeling diode in application. This way, their body diodes exhibit significant reverse recovery charge when commutating from conduction into blocking mode leading to considerable losses [49] (see Sect. 2.1.8). IGBTs have no body diode, and many IGBT modules comprise an additional optimized diode in parallel to achieve a good reverse conduction behavior. GaN transistors do not have a body diode, either. However, when the source node rises above the drain-voltage level and the gate exceeds the drain by one V_{th}, the transistor starts conducting. This is referred to as reverse conduction of the so called "quasi-body diode." In case of a negative gate-source voltage, applied to the GaN during off-state, the source needs to reach $V_{th} + | - V_{GS}|$ above the drain until it starts to conduct leading to an increased forward voltage of the quasi-body diode. For instance, at $V_{GS} = -5$ V and $V_{th} = 1.5$ V, the quasi-body diode has a forward voltage of > 6.5 V. This effect and solutions for the gate driver are described in Sect. 2.3.2. As there is no real body diode, GaN transistors have no reverse recovery charge [14], which is a significant advantage compared to SJMOSFETs [74]. Fast switching applications which benefit from low reverse recovery charge are typical candidates for the implementation of GaN switches.

Dynamic $R_{DS,on}$ Trapped electrons with different time constants in a GaN transistor lead to an increased $R_{DS,on}$, the so called current collapse. This effect slowly decreases as long as the traps exist resulting in a dynamic $R_{DS,on}$ [57]. Most leading manufacturers claim that the dynamic $R_{DS,on}$ is no substantial issue anymore in their devices [57]. Although this effect is significantly reduced, it might still have an impact on the safe-operating area and losses in the application [57]. Panasonic stated to be free of this effect [59].

2.2.3.2   Safe-Operating Area (SOA)

Avalanche Capability   GaN has no avalanche capability, compared to an IGBT and SJMOSFET. Hence, SJMOSFET and IGBTs are more suitable in circuits with unclamped inductive energy [74, p. 44]. As an overvoltage at a GaN transistor usually leads to its destruction, the selected device needs to have a sufficient margin towards its breakdown voltage.

Short Circuit Resistance   In safety applications, the power transistors often need to withstand a short-circuit condition for a specified time. Conventional specifications for silicon devices demand a shut-off of the power transistors by an auxiliary circuit within ~10 ms in case of a short circuit [75, 76]. Although GaN transistors are very small and their cooling is very challenging, [76–78] show a highly robust GaN device (p-gate structure) along with a gate driver solution, capable to meet the conventional requirements in case of a short circuit.

Operation Temperature   GaN can handle higher temperatures compared to silicon, which can be beneficial in some applications. It is a recent research and development topic, to fully exploit assembly methods and peripheral circuits, like the gate drivers, that can withstand these high temperatures as well [79]. In Chap. 5, an approach is proposed for placing the gate driver in large distance to the power transistor.

Device Cooling   GaN transistors dissipate lower power. Nevertheless, the cooling for GaN transistors is more challenging. The dissipated power concentrates on extremely small dies and packages of GaN. Less area for heat sink attachment is available [80]. Insufficient cooling limits the safe-operating area (SOA) of the device. Section 2.3.2 introduces a three-level gate drive scheme, contributing to a more efficient GaN transistor operation. In Chap. 4, a gate driver is presented that supports a three-level gate drive scheme.

2.2.3.3   Gate Control

Gate Voltage   Figure 2.11 compares typical gate-source voltages during power transistor on-state. Silicon transistors require a $V_{GS} \sim 10$ V during on-state while 3.5 V to 6 V are sufficient for GaN, which further lowers the required gate charge. Depending on the gate structure, GaN transistors have very tight gate-voltage ratings, such as EPC transistors that require a $V_{GS} \sim 5$ V during on-state but are limited from -4 V to 6 V. Silicon devices have a threshold voltage V_{th} of ~3 V which is about two times higher than that of GaN transistors. This increases the risk of an unintended GaN turn-on during off-state, which is further described in Sect. 2.3.1.

Switching Speed GaN transistors support fast switching frequencies and fast switching transitions. This requires gate drivers to support a fast gate charge/discharge speed. The fast switching transitions are challenging for the driver to keep the transistor in off-state, which is further explained in Sect. 2.3.1. Generally, GaN devices require lower gate drive currents to achieve fast switching, since they exhibit significantly lower gate charge compared to their silicon counterparts.

GaN gate injection transistor (GIT) According to Sect. 2.2.2, the gate structure of GaN-GIT devices show a diode behavior between the gate and source nodes (Fig. 2.10). This brings the advantage of a clamping behavior preventing a gate overshoot (see Sect. 2.1.4). However, during on-state, the gate driver needs to deliver a constant gate current and typically, when driving GaN GIT transistors, the gate driver needs a current limitation during on-state (Fig. 2.10). Nevertheless, it still needs a low-resistance connection to the driver supply for fast turn-on and to buffer interferences into the gate. Conventionally, this is achieved by a capacitor in parallel to a current limiting resistor in the gate path [81]. Section 4.2.4 presents an approach for the gate drivers in Chaps. 4 and 5 that support both types of GaN transistors, GIT and non-GIT.

Monolithic Gate Driver Integration GaN transistors are lateral devices (see Sect. 2.2.2) in contrast to silicon devices, which makes it more easier to integrate additional circuitry (e.g., gate driver) on the same die with the power transistor. Section 2.1.4 elaborates on the advantages and disadvantages of a monolithic gate driver integration.

2.2.3.4   Applications

The different characteristics of GaN transistors compared to silicon devices have an impact on where and how to implement them in the application.

GaN transistors benefit currently in high-end and high switching frequency solutions. Light detection and ranging (LiDAR), for instance, requires accurate 1 ns current pulses in the 100 A-range at voltages near 100 V with fast repetition rate for driving a laser, in a compact design [82]. GaN transistors achieve a 10 times faster switching, higher efficiency, and smaller footprint compared to their silicon counterpart, enabling a more detailed and faster scanning of the surrounding area [83]. This opens up new fields of application, like real-time motion detection for video gaming, computers that react to hand movements, and autonomous vehicles [84]. Highly compact and efficient DCDC converters are a further emerging application for GaN transistors that are required to convert vehicle board net voltages from 48 V at high switching frequencies into voltages below 5 V to supply various control units [85]. Especially in applications with the importance of low reverse conduction losses, GaN is superior to SJMOSFETs [74]. An example is the so called totem-pole power-factor correction (PFC) [86–88]). In that circuit topology, GaN is typically used in the fast switching half-bridge, while low-ohmic silicon transistors rectify the grid voltage at low switching frequency.

Currently, GaN transistors are typically implemented in applications which are in the middle-voltage range of about $100\,V - 1200\,V$ [32, 68, 89]. However, there are already GaN transistors available for a much wider voltage range of $\sim10\,V$ and beyond $1200\,V$ [57, 90–92]. The current ratings range from $\sim0.5\,A$ up to $\sim120\,A$ [91, 93]. SJMOSFETs operate in a similar range of $200\,V - 900\,V$. This work focuses on applications for $\leq 400\,V$, $\leq 10\,A$, which represent current high-volume markets.

There are different opinions if GaN devices lead to a system cost reduction [89, 94, 95]. Silicon transistors still dominate the market, are highly proven in terms of reliability, show robustness, e.g., due to its avalanche capability, often provide sufficient efficiencies, and are cost-effective. However, the cost of GaN transistors seem to further decrease [89, 95] and the performance is expected to improve rapidly. The targeted use of GaN enables a coexistence of GaN and silicon, in the same application where appropriate, which might be a key factor for cost-efficient and high-performing applications. To push the borders of performance and cost reduction, further research and development are necessary, coping with the challenges in application, not only concerning the device but also in terms of implementation and control of GaN transistors.

2.3   Gate Drive Schemes

The way of controlling the gate of the power transistor has significant impact on its switching behavior and reliable operation. This section gives an introduction on various gate drive voltage schemes.

2.3.1   Unipolar and Bipolar Gate Drive Scheme

Unipolar gate control is the most widespread gate drive scheme. It comprises only gate-source voltages greater than or equal to zero, i.e., $V_{GS} = 0\,V$ during off-state and $V_{GS} > V_{th}$ at turn-on. Typical gate-source voltages for various power devices are mentioned in Sect. 2.2.3. For controlling depletion mode devices (see Sect. 2.2.2), a unipolar gate control needs to provide a negative V_{GS}. Figure 2.12 shows a typical half-bridge configuration with two GaN transistors with the corresponding waveforms of the switching node S and three kinds of V_{GS} gate drive schemes. In this example, a soft-switching condition at high-side transistor turn-on is achieved, and a hard-switching condition at falling switching node transition of node S (see Sect. 2.1.8). In the following, the impact of hard-switching is discussed.

The relationship between the gate signal and soft-switching is further described in Sect. 2.3.2.

In case of a unipolar gate control, the hard-switching steep transition of S can cause an unintended high-side driver turn-on. This is due to coupling via the gate-drain capacitance (Miller coupling) when the low-side driver turns on, causing a peak current into the gate. The peak current causes a voltage drop across the parasitic gate loop inductance and resistance, turning on the high-side power transistor, as the unipolar gate voltage curve in Fig. 2.12 (right) shows. The Miller coupling applies not only to the high-side transistor in a half-bridge but also to the low-side transistor, in case that it experiences a steep voltage transition during off-state. In particular, fast switching power MOSFETs and GaN transistors are affected, which allow very fast switching transients. GaN transistors additionally have a low threshold voltage V_{th} of ~1 V, much lower than silicon (see Sect. 2.2.3). This is usually tackled by placing the gate driver as close as possible next to the power transistor, keeping the gate loop inductance as small as possible. In addition, applying a negative gate voltage enhances the safety margin towards V_{th}, resulting in a bipolar gate-driving scheme Fig. 2.12 (right). Besides improved immunity against coupling, [2] mentions a faster turn-off of IGBTs with a bipolar gate drive scheme. However, as a disadvantage, the bipolar gate drive scheme causes more losses in a GaN transistor when operated in reverse conduction, which is the case in soft-switching condition. This is discussed in more detail in Sect. 2.3.2.

Conventional gate drivers with bipolar gate voltages (Fig. 2.13) consist often of a unipolar gate driver IC with the source of the GaN transistor connected to an intermediate voltage level. Two external capacitors buffer the gate charge for the

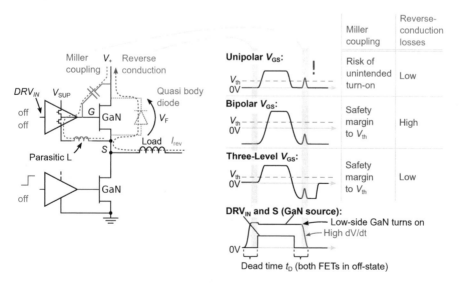

Fig. 2.12 Evaluation of unipolar, bipolar, and three-level gate drive schemes regarding switching robustness and efficiency

Fig. 2.13 Gate driver with conventionally generated bipolar gate voltages

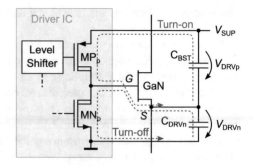

positive and negative voltage levels [81, 96, 97][71, p. 4]. This unipolar gate driver can be a resonant gate driver as well [44].

2.3.2 Multi-level Gate Drive Schemes and Active Gate Control

In many power electronics applications, GaN transistors operate in reverse conduction carrying the inductor current during the dead time t_D when the high-side and low-side switch are off (illustrated at a high-side switch in Fig. 2.12 (left)). As there is no real body diode like in silicon devices, the GaN transistor turns on in reverse operation with a voltage drop V_F across, known as quasi-body diode behavior (see Sect. 2.2.3). A negative gate voltage adds to V_F and significant reverse conduction losses occur at the GaN switch in bipolar gate drive operation [33, 98, 99]. This drawback of a pure bipolar gate drive operation is addressed by a three-level gate drive scheme (positive, 0V, negative gate voltage), as shown in Fig. 2.12 (right), which at the same time provides robustness against unintended turn-on. This was explored for a discrete gate driver in [98, 100].

Besides being robust and reducing the reverse conduction losses at the same time, an adaptable intermediate gate voltage level during transistor turn-on/-off is also used to control the switching speed of the power transistor. Prasobhu et al. [101] controls the gate drive speed by varying the gate-voltage level during GaN turn-on/-off in order to limit the thermal stress of the GaN device to ensure a reliable operation. Other parameters such as electromagnetic interference (EMI) can be controlled with a shaped gate drive voltage [102–104]. For IGBTs, [105] presents a gate driver with three gate voltage levels to achieve a decreased turn-off speed (soft turn-off) in case of a detected short circuit in the IGBT.

2.4   Gate Driver Supply and Signal Transmission

2.4.1   Isolated and Non-isolated Gate Driver Supplies

Gate drivers can be supplied via a galvanically isolated or a non-galvanically isolated gate driver supply (Fig. 2.14). These two types will be called in the following "isolated" and "non-isolated" supplies. Figure 2.14 shows these two types exemplary for a conventional gate driver with two buffer capacitors providing a bipolar gate drive scheme like in Fig. 2.13.

Reasons for the need of an isolated gate driver supply are given in [106, 107]. They include the following topics:

- Protecting the control unit and user, respectively, against high-voltage, in case of a failure
- Reducing common mode noise between the control part and the high-voltage domain, with the risk of faulty switching of the power transistor (see Sect. 2.4.5)
- More flexibility of the overall circuit configuration due to separate ground potentials

For the same reasons, a galvanic isolation may also be applied to a low-side transistor. Furthermore, the isolated low-side driver supply ensures that the high-side and low-side gate driver paths are symmetrical, which is important for a critical gate-drive timing. Typically, the isolated supply is realized with a transformer, which transfers the energy to the gate driver (Fig. 2.14a). The rectified voltage at the transformer secondary side is referred to the driver chip ground. These transformers can be both integrated and discrete components. There are other types of isolated driver supply such as an optical power transducer [108].

Figure 2.14b shows a non-isolated driver supply, i.e., no galvanic isolation decouples the driver IC from its supply source V_{SUP}. In case of a low-side driver, it

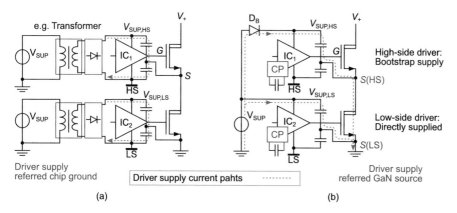

Fig. 2.14 (**a**) Isolated and (**b**) non-isolated gate driver supply in a GaN half-bridge configuration

is simply a direct connection from the source V_{SUP} to the gate driver IC. In case of a high-side driver, the IC is typically supplied by a bootstrap circuit. Other concepts for a non-isolated high-side driver supply often require high-voltage transistors, like the charge pump-based concepts in [109, 110] or the circuits with a series regulator in [111, 112]. Figure 2.14b shows the basic function of a bootstrap circuit, which is the most widespread non-isolated high-side gate driver supply. During half bridge on-state (node $S(HS)$ is connected to V_+), the driver IC is supplied from the buffer capacitors. The bootstrap diode D_{B1} prevents a discharge of the buffer capacitors. If the half bridge is in off-state, the buffer capacitors are recharged over the bootstrap diode D_B and the conducting low-side power transistor.

As described in Sect. 2.3, a negative gate voltage supply is a very effective method to obtain a robust gate driver operation. The current paths in Fig. 2.14 show that the buffer capacitor buffering the negative gate voltage can only be recharged using an isolated driver supply. In case of a non-isolated driver supply, the current loop is always closed over the source terminals of the power transistors. Hence, the gate driver may require an additional charge pump to charge the buffer capacitor for the negative gate bias [113]. The charge pump needs to deliver the whole power-transistor gate charge for every transistor turn-off event comprising typically an area consuming integrated [21] or an off-chip charge pump capacitor. The discrete bipolar drivers in [71, p. 3] and [97] require no additional charge pump, but still need two external capacitors. Nagai et al. [114] presents a microwave-driven bipolar gate driver without any external capacitors; however, it does not buffer the transmitted energy, limiting the gate current to a low value of $\sim$30 mA. Another solution requires the implementation of an additional bootstrap supply referred to V_+ instead of the application ground [66, 115]. Since this supply can be used for multiple high-side transistors, its benefit increases for a larger number of high-side transistors (e.g., bridges in motor control). The three-level gate driver presented in Chap. 4 [34] can be supplied with both an isolated and a non-isolated driver supply with fully integrated buffer capacitors, and it does not require an additional charge pump capacitor.

2.4.2   Bootstrap Driver Supply

The basic function of the bootstrap circuit is explained in Sect. 2.4.1. The bootstrap circuit is a very compact and low cost option for supplying a high-side gate driver. Therefore, it is commonly applied in many designs. Its limitations and challenges in various applications are considered in this section.

Limited Support of Bipolar Gate Drive Voltage A bootstrap circuit typically does not support gate drivers with bipolar gate drive voltages, as described in Sect. 2.4.1.

Duty Cycle Restrictions The recharge current of the bootstrap capacitor C_{DRV} is typically limited, e.g., by a resistor, to prevent oscillations or an overcharging of

C_{DRV}, which is discussed further below in this section. Therefore, the bootstrap circuit requires a minimum half-bridge off-state duration.

During half-bridge on-state, the high-side gate driver is supplied from C_{DRV}. The size of C_{DRV} and the power consumption of the gate driver during on-state limit the maximum half-bridge on-time. These timing restrictions result in a minimum and maximum allowed duty cycle for the high-side transistor. To achieve 100 % power transistor on-state duration, Seidel et al. [116] presents a solution with its principle explained in Sect. 2.4.4. Lin and Lee [110] proposes a transformerless charge pump-based solution that requires high-voltage transistors.

Overcharging of the Bootstrap Capacitor   When supplying the gate driver with a bootstrap circuit, particular precautions in the design must be considered to prevent an overvoltage at the bootstrap capacitor [117–119]. Figure 2.15 shows a typical half-bridge application. After the half-bridge turns off (low-side transistor turns on), the source S of the high-side transistor may drop below ground due to ringing caused by parasitic inductances. This undershoot can achieve values of about -10 V. Also during dead time (both transistors are off), the low-side transistor is in reverse conduction, S drops below ground by the forward voltage of the body diode. This voltage drop adds to the bootstrap supply voltage (see Fig. 2.15). Without precautions, an excessive charging of the bootstrap capacitor occurs [117]. This is in particular critical to applications with GaN transistors because of two reasons:

1. The tight gate voltage maximum ratings of GaN allow only a small overvoltage at the gate [12, 33] (see Sect. 2.2.3).
2. GaN transistors have a quasi-body diode behavior with a large voltage drop across their drain-source terminals during reverse conduction as described in Sects. 2.2.3 and 2.3. Therefore, the charging voltage of C_{DRV} is higher.

There are several solutions. An external Schottky diode ($V_F \sim 0.3$ V) can be implemented in parallel to the drain-source terminals of the power transistor limiting the negative swing of the source S [12, 120]. However, the diode needs to be capable of carrying the whole load current, and its junction capacitance adds to the parasitic drain-source capacitance of the power device [120]. Furthermore, an overvoltage at the bootstrap capacitor can be clamped by a Zener diode at the expense of a significant increase of the gate driver power consumption [120]. A linear regulator (LDO), providing a regulated gate driver supply from the unregulated bootstrap capacitor voltage, provides a better efficiency and enables to adjust the gate driver supply voltage [12, 121, 122]. This solution typically requires two buffer capacitors: the bootstrap capacitor and a capacitor at the LDO output, buffering the gate driver supply. There are several concepts of filtering the transient currents that might overcharge the bootstrap capacitor. The value of the bootstrap capacitor can be increased [120], which contradicts to an on-chip integration of the bootstrap capacitor. Moreover, also a discrete capacitor can only be increased to a limited size [120]. A simple and widely used approach to filter the transient currents is to insert a resistor in the bootstrap capacitor recharge path. Roschatt et al. [120] presents an inductor instead of a resistor in the recharge path, but this also prevents a

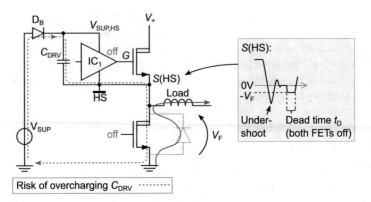

Fig. 2.15 Bootstrap circuit supplying a high-side gate driver

fully integrated gate driver. All filtering concepts suffer from an increased capacitor recharge time, limiting the minimum allowed duty cycle of the half bridge. In particular for a GaN driver, the fully integrated approach in [119] detects node S(HS) and starts the recharging of the bootstrap capacitor as soon as S(HS) reaches 0 V after its negative swing. This approach provides a shorter capacitor recharge time, limiting the minimum off-time of the half-bridge. The gate driver based on HVES, described in Chap. 4, is not very sensitive to driver supply voltage variations, relaxing the demand on precautions against a C_{DRV} overvoltage (see Sect. 2.4.2).

Start-Up In order to charge the bootstrap capacitor, the half-bridge low-side transistor needs to be in the on-state (see Fig. 2.15). Depending on the application, a start-up procedure in the control unit can ensure that the low-side transistor is switched on before the bootstrap supplied high-side driver enters normal operation. Seidel et al. [116] presents a solution with an additional small transformer-based supply that can charge the bootstrap capacitor during start-up, further explained in Sect. 2.4.4.

Coupling Currents and Reverse Recovery in the Bootstrap Diode During half-bridge switching, steep slopes occur at the floating ground potential of the high-side gate driver, referred to the application ground, which is further described in Sect. 2.4.5. Any parasitic capacitances from the high-side to the low-side lead to coupling currents, which cause losses and even malfunctions in the application. Therefore, a bootstrap diode D_B with a very low junction capacitance is crucial. Furthermore, the reverse recovery charge of D_B should be very low to avoid significant reverse currents during the rising edge of S(HS). The reverse recovery currents discharge C_{DRV} decreasing the bootstrap supply efficiency.

2.4.3 Signal Transmission

Similar to the gate driver supply, the signal transmission for the gate driver turn-on and -off signals can be galvanically isolated or not isolated. In addition to the gate control signals, there are other signals to be transmitted, like error messages, status signals, feedback signals for driver supply control [116, 123].

There are various ways for a signal transmission, listed below:

- **Magnetically:** Pulsed signals via transformer [124, 125], frequency shift keying modulation via transformer [116], or transmitting coil with receiving GMR (giant magnetoresistance) [107].
- **Optically:** Via optocoupler, fiber optics, or optical transducer [106, 126, 127].
- **Capacitively:** Capacitively coupled gate drivers use various modulation schemes and transmit signals typically differentially via two capacitors [128–130].
- **Piezoelectric transformer:** A signal transmission for gate drivers via piezoelectric transformer is still under research [130, 131].
- **Non-isolated:** Via level shifter [132].

There are many performance indicators of the different signal transmission methods, like propagation delay, delay matching between two transmission channels, reliability, isolation capability, robustness against common-mode transient immunity (CMTI), etc. A comparison of the main topologies is given in [107, 133], and for some of them also in [2, 106].

2.4.4 Combined Driver Supply and Signal Transmission

Some solutions share a single isolation barrier for energy and signal transmission for controlling and supplying the gate driver. Seidel et al. [116] presents a solution with a bootstrap supply circuit to provide the main switching energy for the power transistor. A signal transformer transmits the turn-on and turn-off signals. In parallel to the bootstrap supply, this small signal transformer delivers an additional power in the 10 mW-range. This overcomes some duty cycle restrictions of a pure bootstrap driver supply, e.g., it enables a 100 % power transistor on-state duration (see Sect. 2.4.2). Additionally, a bidirectional signal transmission is realized via the signal transformer. This enables a very compact solution. As GaN transistors require lower gate charges, the energy transfer via the small signal transformer might be sufficient. The absence of a bootstrap diode reduces the parasitic coupling capacitance between the low-side and high-side, decreasing common mode currents during voltage transitions (see Sect. 2.4.5).

In [114], a rectified microwave turns on a GaN transistor. Shutting off the microwave signal discharges the GaN gate. This possibility delivers gate currents of only $\sim$20 mA. Nagai et al. [134] stores the energy from the microwave in an external capacitor achieving peak gate currents of 170 mA with GaN and

2 A with an IGBT. The required gate current for the GaN transistor is much lower, because its current capability is chosen to be much lower than that of the IGBT. Furthermore, GaN devices generally require lower gate currents to achieve fast switching (see Sect. 2.2.3). However, the measurements in [114, 134] exhibit relatively slow switching slew rates for GaN of less than $5\,\mathrm{V\,ns^{-1}}$. As this is an emerging technology, it should be further considered, how the systems in [114, 134] behave for faster transitions.

2.4.5 Coupling Currents Between Low-Side and High-Side

Figure 2.16 shows a half-bridge application with the gate driver control signals and the power supply transferred via a high-voltage barrier, which can be isolated or non-isolated (see Sect. 2.4.1). During the voltage transittiions of the switching node $S(\mathrm{HS})$, any parasitic coupling capacitances (accounted for by C_{cpl}) of the high-voltage barrier between the low-side and high-side need to be charged and discharged. This results in common mode coupling currents with their paths depicted in Fig. 2.16 for a typical case. In applications with GaN transistors, extremely steep voltage transitions of several $100\,\mathrm{V\,ns^{-1}}$ occur (see also Sect. 2.2.3). For example, if $S(\mathrm{HS})$ rises with $200\,\mathrm{V\,ns^{-1}}$ and assuming a low value $C_{\mathrm{cpl}} = 1\,\mathrm{pF}$, a coupling current of $200\,\mathrm{mA}$ $(1\,\mathrm{pF} \cdot 200\,\mathrm{V\,ns^{-1}})$ flows. This confirms that a low overall coupling capacitance C_{cpl} is crucial. Furthermore, this has major influence on robust signal transmission and detection. The coupling currents charge and discharge also C_{DRV}, which need to be considered in gate driver designs with a small C_{DRV}, as is the case in this work. Section 4.2.3 further describes and addresses this issue.

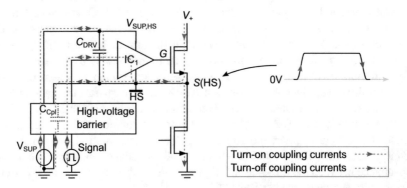

Fig. 2.16 Typical coupling currents controlling a high-side transistor in a half-bridge configuration

Appendix A: Exemplary Gate Loop Inductance Values

Figure 2.17 shows a typical gate driver (see Fig. 2.2) with exemplary inductance values of the interconnections between buffer capacitor C_{DRV}, gate driver, gate resistor R_G, and power transistor listed in the table. This set of inductance values is given for four different gate driver implementation scenarios, (1) gate driver in a small-outline integrated circuit (SOIC) package, (2) gate driver in a QFN package, (3) gate driver and power transistor as bare die, and (4) gate driver and power transistor as bare die with direct bonds between driver and transistor without R_G in the gate loop. For C_{DRV} and R_G, a surface-mounted device (SMD) package (size 0603) is assumed, both with ~1 nH parasitic inductance [135]. For the power transistor, a total parasitic inductance of 1 nH for the gate-source path in the transistor package is assumed. The value is exemplary for high-performing packages of modern GaN transistors which is typically assembled in a flip-chip configuration, i.e., it has no bond wires anymore, e.g., [11].

Scenario (1)   For each connection of the gate driver in an SOIC package, an average inductance value of ~3 nH including bond wire is defined based on [136, 137]. Since the pins carrying the gate drive current are typically chosen to be low inductive for each driver connection, an inductance value of only 2.5 nH is assumed.

Scenario (2)   A QFN package, also typical for gate drivers, reduces the average bond wire inductance significantly to ~0.5 nH [136]. This package has no lead frame and an exposed ground pad leading to a small loop formed by the driver output connection and the ground return path resulting in a small parasitic inductance.

Scenarios (3) and (4)   The gate driver and transistor implemented as bare die. This allows for shorter bond wires at the gate driver. However, the transistor now also needs bond wires, increasing the parasitic inductances. Based on [13], a parasitic trace inductance of 3 nH is chosen for the wire paths between the gate driver and transistor. This value is 0 for scenario 4 because the direct bonds between the gate driver and the transistor avoid substrate connections. For the wires on substrate between C_{DRV} and the gate driver, 0.5 nH are assumed. This value increases significantly if C_{DRV} cannot be implemented near to the driver package. For example, the placing of C_{DRV} at the opposite side of the gate driver requires

Exemplary inductances (in nH) of	1)	2)	3)	4)
Capacitor C_{DRV} (SMD 0603)	1	1	1	1
Resistor R_G (SMD 0603)	1	1	1	0
Average gate driver connection (⌇)	2.5	0.5	1.5	1.5
Average transistor connection (⌇)	0.5	0.5	1.5	1.5
External traces: Driver - transistor	3	3	3	0
External traces: C_{DRV} - driver	0.5	0.5	0.5	0.5

1) SOIC driver package, 2) QFN driver package, 3) Bare die driver & transistor,
4) Bare die driver & transistor, without R_G (driver & transistor directly bonded)

Fig. 2.17 Typical inductance values of elements in the gate loop

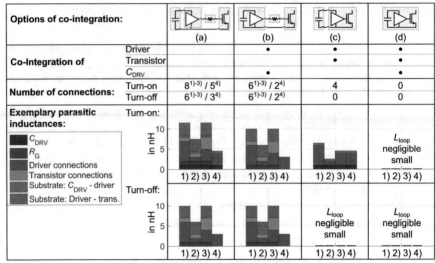

Options of co-integration:		(a)	(b)	(c)	(d)
Co-Integration of	Driver		•	•	•
	Transistor			•	•
	C_{DRV}			•		•
Number of connections:	Turn-on	$8^{1)-3)} / 5^{4)}$	$6^{1)-3)} / 2^{4)}$	4	0
	Turn-off	$6^{1)-3)} / 3^{4)}$	$6^{1)-3)} / 2^{4)}$	0	0

Exemplary parasitic inductances:

- C_{DRV}
- R_G
- Driver connections
- Transistor connections
- Substrate: C_{DRV} - driver
- Substrate: Driver - trans.

Fig. 2.18 Exemplary gate loop inductance values and number of interconnections assuming gate drive integration options (a) to (d) of buffer capacitor, gate driver, and power transistor

two via connections through the printed circuit board (PCB) with each ∼1 nH [138]. On substrates with only a single wiring layer, which is often the case for ceramic substrates, the substrate inductance between C_{DRV} and the gate driver can be significantly higher [34].

Figure 2.18 shows various integration options of the gate driver, buffer capacitor C_{DRV}, and the power transistor, same as Fig. 2.5. However, Fig. 2.18 shows the exemplary gate loop inductance values not only for scenarios (2) and (4), but also for scenarios (1) (SOIC driver package) and (3) (driver and transistor as bare dies with external R_G).

Appendix B: Losses Caused by C_{oss}

As explained in Sects. 2.1.8 and 2.1.7, the losses caused by the output capacitance C_{oss} of a power transistor depend on whether C_{oss} is charged/discharged via a loss dissipating element like a resistor or transistor channel, or resonantly. Figure 2.19a shows a QV-diagram (see Sect. 2.1.7) of a linear C_{oss}, i.e., C_{oss} is independent from V_{DS}. If C_{oss} is charged via a power-dissipating path during power transistor off-state, the area under the QV-curve indicates the dissipated energy. When the power transistor turns on, the transistor channel dissipates the stored energy in C_{oss}. The triangular green area above the QV-curve is a measure for the dissipated energy.

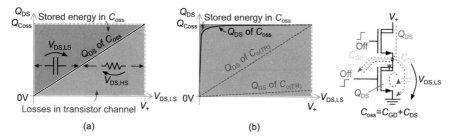

Fig. 2.19 QV-diagrams and the equivalent circuit for charging (**a**) a linear (i.e., voltage-independent) C_{oss} and (**b**) a highly non-linear C_{oss}

For a linear C_{oss}, the same amount of losses occurs for both, turning on and off the power transistor.

Figure 2.19b shows a QV-diagram of a highly non-linear C_{oss}, like it is the case for the newest CoolMOS^{TM} generation C7 [139]. This non-linearity is a design goal, since it minimizes the C_{oss} losses during power transistor turn-on. Furthermore, it leads to fast V_{DS} transitions and a quasi-zero-voltage switching at transistor turn-off [139] which results in significantly lower transition losses (see Sect. 2.1.8). The dissipated energy E_{oss} caused by C_{oss} during power transistor turn-on can be calculated with the aid of the datasheet value $C_{o(ER)}$ (effective energy-related output capacitance). $C_{o(ER)}$ stores the same energy as C_{oss} when V_{DS} is charged up to the final V_{DS} (V_+) [11]. Hence, E_{oss} is defined as

$$E_{oss} = 0.5 \cdot C_{o(ER)} \cdot V_+{}^2. \tag{2.7}$$

During low-side transistor off-state, C_{oss} is charged. In several applications, such as a typical boost converter, C_{oss} is charged via an inductor. Hence, the energy under the QV-curve is not dissipated. However, this is not the case for a typical hard-switching half-bridge configuration, as shown in Fig. 2.19 (right). The energy $E_{chargingLossCoss}$ is dissipated in the high-side transistor channel when charging C_{oss}. $E_{chargingLossCoss}$ can be calculated with the datasheet value $C_{o(TR)}$ (effective time-related output capacitance), which gives the same charging time as C_{oss} while charging it via a constant current from 0 V to the final V_{DS} (V_+) [11]:

$$E_{chargingLossCoss} = Q_{Coss} \cdot V_+ - E_{oss} = (C_{o(TR)} - 0.5 \cdot C_{o(ER)}) \cdot V_+{}^2. \tag{2.8}$$

$E_{chargingLossCoss}$ can be significantly higher than the dissipated energy E_{oss} during transistor turn-on depending on the power transistor. Table 2.1 shows an example of a hard-switched modern SJMOSFET IPP65R065C7 that causes a C_{oss} discharging loss of only 8 µJ which is only marginally higher compared to the GaN transistor GS66508T causing 7 µJ. However, for the silicon SJMOSFET, the C_{oss} charging losses of 170 µJ are a factor of ∼10 higher than that of the GaN transistor. Hence,

Table 2.1 Output capacitance comparison of a GaN and silicon transistor

	Si CoolMOS IPP65R065C7 (700 V, 33 A)	GaN GS66508T (650 V, 30 A)
C_{oss} ($C_{GD} + C_{DS}$)	48 pF	65 pF
$C_{o(ER)}$	100 pF	88 pF
$C_{o(TR)}$	1110 pF	142 pF
E_{oss} (@ $V_+ = 400$ V)	8 μJ	7 μJ
$E_{chargingLossCoss}$ (@ $V_+ = 400$ V)	170 μJ	16 μJ

the total silicon losses caused by C_{oss} are by a factor of ∼8 higher although the SJMOSFET has a lower C_{oss} datasheet value compared to the GaN transistor.

References

1. Wicht, B., Wittmann, J., Seidel, A., & Schindler, A. (2016). Wideband continuous-time $\sum \Delta$ ADCs, Automotive electronics, and power management – advances in analog circuit design 2016. In: A. Baschirotto, P. Harpe, & K. A. A. Makinwa, *High-voltage fast-switching gate drivers*. Berlin, Heidelberg: Springer. Chap. 9. ISBN: 978-3-319-41670-0.
2. Herzer, R. (2010, March). Integrated gate driver circuit solutions. In: *2010 6th International Conference on Integrated Power Electronics Systems (CIPS)* (pp. 1–10).
3. Seidel, A., Costa, M. S., Joos, J., & Wicht, B. (2015). Area efficient integrated gate drivers based on high-voltage charge storing. In: *IEEE Journal of Solid-State Circuits, 50*(7), 1550–1559. ISSN: 0018-9200. https://doi.org/10.1109/JSSC.2015.2410797
4. Xue, Y., Wang, Z., Tolbert, L. M., & Blalock, B. J. (2013, June). Analysis and optimization of buffer circuits in high current gate drives. In: *Transportation Electrification Conference and Expo (ITEC), 2013 IEEE* (pp. 1–6). https://doi.org/10.1109/ITEC.2013.6574495
5. Ng, J. C. W., Trescases, O., & Wei, G. (2007, December). Output stages for integrated DC-DC converters and power ICs. In: *Proceedings of IEEE Conference on Electron Devices and Solid-State Circuits, 2007. EDSSC 2007* pp. 91–94. https://doi.org/10.1109/EDSSC.2007.4450069.
6. Huque, M. A., Vijayaraghavan, R., Zhang, M., Blalock, B. J., Tolbert, L. M., & Islam, S. K. (2007, June). An SOI-based high-voltage, high-temperature gate-driver for SiC FET. In: *Power Electronics Specialists Conference, 2007. PESC 2007. IEEE* (pp. 1491–1495). https://doi.org/10.1109/PESC.2007.4342215
7. Greenwell, R. L., McCue, B. M., Tolbert, L. M., Blalock, B. J., & Islam, S. K. (2013, March). High-temperature SOI-based gate driver IC for WBG power switches. In: *Applied Power Electronics Conference and Exposition (APEC), 2013 Twenty-Eighth Annual IEEE* (pp. 1768–1775). https://doi.org/10.1109/APEC.2013.6520535
8. *Design Guide for Selection of Bootstrap Components* (2008, October). Fairchild Semiconductor Corporation, Appl. Note AN-9052.
9. Kang, T., Lee, Y., Park, M., & Kim, J. (2013, November). A 15-V 40-kHz class-D gate driver IC with 62% energy recycling rate. In: *2013 IEEE Asian Solid-State Circuits Conference (A-SSCC)* (pp. 377–380). https://doi.org/10.1109/ASSCC.2013.6691061
10. Xu, J., Sheng, L., & Dong, X. (2012, September). A novel high speed and high current FET driver with floating ground and integrated charge pump. In: *Energy Conversion Congress and Exposition (ECCE), 2012 IEEE* (pp. 2604–2609). https://doi.org/10.1109/ECCE.2012.6342393

11. *GS66508T, Top-Side Cooled 650 V E-Mode GaN Transistor, Preliminary Datasheet* (2018). GaN Systems Incorporation.
12. Lidow, A. (2014). Driving GaN transistors. In: *GaN transistors for efficient power conversion.* Power Conversion Publications. ISBN: 978-1-118-84476-2.
13. Acuña, J., Seidel, A., & Kallfass, I. (2017, December). Design and implementation of a gallium-nitride-based power module for light electro-mobility applications. In: *Proceedings of IEEE Southern Power Electronics Conference (SPEC)* (pp. 1–6). https://doi.org/10.1109/SPEC.2017.8333625
14. Reusch, D., Gilham, D., Su, Y., & Lee, F. C. (2012, February). Gallium nitride based 3D integrated non-isolated point of load module. In: *Proceedings of Twenty-Seventh Annual IEEE Applied Power Electronics Conference and Exposition (APEC)* (pp. 38–45). https://doi.org/10.1109/APEC.2012.6165796
15. Hughes, B., Chu, R., Lazar, J., Hulsey, S., Garrido, A., Zehnder, D., et al. (2013, October). Normally-off GaN switching 400V in 1.4ns using an ultra-low resistance and inductance gate drive. In: *Proceedings of 1st IEEE Workshop Wide Bandgap Power Devices and Applications* (pp. 76–79). https://doi.org/10.1109/WiPDA.2013.6695566
16. Mishra, D., Arora, V., Nguyen, L., Iriguchi, S., Sada, H., Clemente, L., et al. (2017, May). Packaging innovations for high voltage (HV) GaN technology. In: *Proceedings of IEEE 67th Electronic Components and Technology Conference (ECTC)* (pp. 1480–1484). https://doi.org/10.1109/ECTC.2017.322
17. Simonot, T., Rouger, N., & Crebier, J. (2010, September). Design and characterization of an integrated CMOS gate driver for vertical power MOSFETs. In: *Proceedings of IEEE Energy Conversion Congress and Exposition* (pp. 2206–2213). https://doi.org/10.1109/ECCE.2010.5617824
18. Fichtenbaum, N., Giandalia, M., Sharma, S., & Zhang, J. (2017, September). Half-bridge GaN power ICs: Performance and application. *IEEE Power Electronics Magazine, 4*(3), pp. 33–40. Navitas driver. ISSN: 2329-9207. https://doi.org/10.1109/MPEL.2017.2719220
19. *LMG3410 600-V 12-A Integrated GaN Power Stage* (2017, April). Texas Instruments Incorporated.
20. Semiconductor, Dialog (2016). *DA8801 SmartGaN™Integrated 650V GaN Half Bridge Power IC*. Dialog Semiconductor. https://www.dialog-semiconductor.com/sites/default/files/da8801_smartgan_product_brief.pdf
21. Rose, M., Wen, Y., Fernandes, R., Van Otten, R., Bergveld, H. J., & Trescases, O. (2015, May). A GaN HEMT driver IC with programmable slew rate and monolithic negative gate-drive supply and digital current-mode control. In: *Proceedings of IEEE 27th International Symposium on Power Semiconductor Devices IC's (ISPSD)* (pp. 361–364). https://doi.org/10.1109/ISPSD.2015.7123464
22. *EPC2112 – 200 V, 10 A Integrated Gate Driver eGaN® IC – Preliminary Datasheet* (2018, March). Efficient Power Conversion Corporation.
23. Moench, S., Costa, M., Barner, A., Kallfass, I., Reiner, R., Weiss, B., et al. (2015, November). Monolithic integrated quasi-normally-off gate driver and 600 V GaN-on-Si HEMT. In: *Proceedings of IEEE 3rd Workshop Wide Bandgap Power Devices and Applications (WiPDA)* (pp. 92–97). https://doi.org/10.1109/WiPDA.2015.7369264
24. Moench, S., Reiner, R., Weiss, B., Waltereit, P., Quay, R., Kaden, T., et al. (2018, June). Towards highly-integrated high-voltage multi-MHz GaN-on-Si power ICs and modules. In: *Proceedings of Renewable Energy and Energy Management PCIM Europe 2018; International Exhibition and Conference for Power Electronics, Intelligent Motion* (pp. 1–8).
25. Zhu, M., & Matioli, E. (2018, May). Monolithic integration of GaN-based NMOS digital logic gate circuits with E-mode power GaN MOSHEMTs. In: *Proceedings of IEEE 30th International Symposium on Power Semiconductor Devices and ICs (ISPSD)* (pp. 236–239). https://doi.org/10.1109/ISPSD.2018.8393646
26. Ujita, S., Kinoshita, Y., Umeda, H., Morita, T., Kaibara, K., Tamura, S., et al. (2016, June). A fully integrated GaN-based power IC including gate drivers for high-efficiency DC-DC converters. In: *Proceedings of IEEE Symposium on VLSI Circuits (VLSI-Circuits)* (pp. 1–2). https://doi.org/10.1109/VLSIC.2016.7573496

27. Tang, G., Kwan, M.-H., Zhang, Z., He, J., Lei, J., Su, R.-Y., et al. (2018, May). High-speed, high-reliability GaN power device with integrated gate driver. In: *Proceedings of IEEE 30th International Symposium on Power Semiconductor Devices and ICs (ISPSD)* (pp. 76–79). https://doi.org/10.1109/ISPSD.2018.8393606

28. Kaufmann, M., Seidel, A., & Wicht, B. (2020, March). Long, short, monolithic – The gate loop challenge for GaN drivers: Invited paper. In: *Proceedings of IEEE Custom Integrated Circuits Conference (CICC)* (pp. 1–5). https://doi.org/10.1109/CICC48029.2020.9075937

29. Simonot, T., Crebier, J., Rouger, N., & Gaude, V. (2010, September). 3D hybrid integration and functional interconnection of a power transistor and its gate driver. In: *Proceedings of IEEE Energy Conversion Congress and Exposition* (pp. 1268–1274). https://doi.org/10.1109/ECCE.2010.5617816

30. *Superjunction MOSFET for Charger Applications – 600 V/650 V/700 V/800 V Cool-MOS™CE* (2016, February). Infineon Technologies AG.

31. Kaminski, N., Hilt, O. (2012, March). SiC and GaN devices – Competition or coexistence? In: *Proceedings of 7th International Conference on Integrated Power Electronics Systems (CIPS)* (pp. 1–11).

32. Kaufmann, M., Lueders, M., Kaya, C., & Wicht, B. (2020). 18.2 A monolithic E-mode GaN 15W 400V offline self-supplied hysteretic buck converter with 95.6% efficiency. In: *Proceedings of IEEE International Solid- State Circuits Conference – (ISSCC)* (pp. 288–290).

33. Seidel, A., & Wicht, B. (2017, February). 25.3 A 1.3A gate driver for GaN with fully integrated gate charge buffer capacitor delivering 11nC enabled by high-voltage energy storing. In: *2017 IEEE International Solid-State Circuits Conference (ISSCC)* (pp. 432–433). https://doi.org/10.1109/ISSCC.2017.7870446

34. Seidel, A., & Wicht, B. (2018). Integrated gate drivers based on high-voltage energy storing for GaN transistors. In: *IEEE Journal of Solid-State Circuits, 53*, pp. 1–9. ISSN: 0018-9200. https://doi.org/10.1109/JSSC.2018.2866948

35. Wolfgang, F., & Ziqing, Z. (2018). *Turn-on performance comparison of current-source vs. voltage-source gate drivers*. Infineon Technologies AG. https://www.psma.com/sites/default/files/uploads/tech-forums-safety-compliance/presentations/is041-turn-performance-comparison-current-source-vs-voltage-source-gate-drivers.pdf

36. Schindler, A., Koeppl, B., Pottbaecker, A., Zannoth, M., & Wicht, B. (2016, September). Gate driver with 10/15ns in-transition variable drive current and 60% reduced current dip. In: *Proceedings of ESSCIRC Conference 2016: 42nd European Solid-State Circuits Conference* (pp. 325–328). https://doi.org/10.1109/ESSCIRC.2016.7598308

37. Schindler, A., Koeppl, B., Wicht, B., & Groeger, J. (2017, March). 10ns Variable current gate driver with control loop for optimized gate current timing and level control for in-transition slope shaping. In: *Proceedings of IEEE Applied Power Electronics Conference and Exposition (APEC)* (pp. 3570–3575). https://doi.org/10.1109/APEC.2017.7931210

38. Zhang, Z., Xu, P., & Liu, Y. (2013). Adaptive continuous current source drivers for 1-MHz boost PFC converters. *IEEE Transactions on Power Electronics, 28*(5), 2457–2467. ISSN: 0885-8993. https://doi.org/10.1109/TPEL.2012.2218619

39. Eberle, W., Zhang, Z., Liu, Y., & Sen, P. C. (2008). A current source gate driver achieving switching loss savings and gate energy recovery at 1-MHz. *IEEE Transactions on Power Electronics, 23*(2), 678–691. ISSN: 0885-8993. https://doi.org/10.1109/TPEL.2007.915769

40. Mashhadi, I. A., Soleymani, B., Adib, E., & Farzanehfard, H. (2018). A dual-switch discontinuous current-source gate driver for a narrow on-time buck converter. *IEEE Transactions on Power Electronics, 33*(5), 4215–4223. ISSN: 0885-8993. https://doi.org/10.1109/TPEL.2017.2723240

41. Mashhadi, I. A., Khorasani, R. R., Adib, E., & Farzanehfard, H. (2017). A discontinuous current-source gate driver with gate voltage boosting capability. *IEEE Transactions on Industrial Electronics, 64*(7), 5333–5341. ISSN: 0278-0046. https://doi.org/10.1109/TIE2017.2674626

42. Mashhadi, I. A., Ovaysi, E., Adib, E., & Farzanehfard, H. (2016). A novel current-source gate driver for ultra-low-voltage applications. In: *IEEE Transactions on Industrial Electronics, 63*(8), 4796–4804. ISSN: 0278-0046. https://doi.org/10.1109/TIE.2016.2554539

43. de Vries, I. D. (2002, March). A resonant power MOSFET/IGBT gate driver. In: *Proceedings of APEC. Seventeenth Annual IEEE Applied Power Electronics Conference and Exposition (Cat. No.02CH37335)* (Vol. 1, pp. 179–185). https://doi.org/10.1109/APEC.2002.989245

44. Badawi, N., Knieling, P., & Dieckerhoff, S. (2012, September). High-speed gate driver design for testing and characterizing WBG power transistors. In: *Proceedings of 15th International Power Electronics and Motion Control Conference (EPE/PEMC)* (pp. LS6d.4-1–LS6d.4-6). https://doi.org/10.1109/EPEPEMC.2012.6397499

45. Long, Y., Zhang, W., Costinett, D., Blalock, B. B., & Jenkins, L. L. (2015, March). A high-frequency resonant gate driver for enhancement-mode GaN power devices. In: *Applied Power Electronics Conference and Exposition (APEC), 2015 IEEE* (pp. 1961–1965). https://doi.org/10.1109/APEC.2015.7104616

46. Bathily, M., Allard, B., & Hasbani, F. (2012). A 200-MHz integrated buck converter with resonant gate drivers for an RF power amplifier. *IEEE Transactions on Power Electronics, 27*(2), 610–613. ISSN: 0885-8993. https://doi.org/10.1109/TPEL.2011.2119380

47. van Schuylenbergh, K., & Puers, R. (2009). *Inductive powering – Basic theory and application to biomedical systems*. Berlin Heidelberg: Springer Science & Business Media. ISBN: 978-9-048-12412-1.

48. Hancock, J., Stueckler, F., & Vecino, E. (2013, April). *Cool MOS^{TM} C7: Mastering the Art of Quickness A Technology Description and Design Guide*. Infineon Technologies AG.

49. *Application Note AN-983 – IGBT Characteristics* (2012, July). Infineon Technologies AG.

50. Udrea, F., Deboy G., & Fujihira, T. (2017). Superjunction power devices, history, development, and future prospects. *IEEE Transactions on Electron Devices, 64*(3), 713–727. ISSN: 0018-9383. https://doi.org/10.1109/TED.2017.2658344

51. Zong, Z., Villamor, A., Liao, J., & Lin, H. (2017, August). *IGBT Market and Technology TRENDS 2017*. Yole Développment. https://www.i-micronews.com/category-listing/product/igbt-market-and-technology-trends-2017.html?utm_source=PR&utm_medium=email&utm_campaign=IGBT_MarketStatus_YOLE_August2017

52. Kaminski, N. (2009, September). State of the art and the future of wide band-gap devices. In: *Proceedings of 13th European Conference on Power Electronics and Applications* (pp. 1–9).

53. Matheson, R. (2015, July). Making the new silicon. *MIT News* http://news.mit.edu/2015/gallium-nitride-electronics-silicon-cut-energy-0729

54. Mion, C. (2006). *Investigation of the thermal properties of gallium nitride using the three omega technique*. PhD thesis. North Carolina State University.

55. Lutz, J., Schlangenotto, H., Scheuermann, U., & De Doncker, R. (2011). *Semiconductor power devices. Physics, characteristics, reliability*. Berlin, Heidelberg: Springer Berlin Heidelberg. ISBN:9783642111259.

56. Trew, R. (1997, June). The operation of microwave power amplifiers fabricated from wide bandgap semiconductors. In: *Microwave Symposium Digest, 1997. IEEE MTT-S International* (Vol. 1, pp. 45–48). https://doi.org/10.1109/MWSYM1997.604512

57. Jones, E. A., Wang, F., & Ozpineci, B. (2014, October). Application-based review of GaN HFETs. In: *Proceedings of IEEE Workshop Wide Bandgap Power Devices and Applications* (pp. 24–29). https://doi.org/10.1109/WiPDA.2014.6964617

58. Longobardi, G., Efthymiou, L., & Arnold, M. (2018, November). GaN power devices for electric vehicles state-of-the-art and future perspective. In: *Proceedings of Ship Propulsion and Road Vehicles Int 2018 IEEE International Conference on Electrical Systems for Aircraft, Railway Transportation Electrification Conference (ESARS-ITEC)* (pp. 1–6). https://doi.org/10.1109/ESARSITEC.2018.8607788

59. Corporation, Panasonic (2013, March). *GaN Power Devices*. Panasonic Corporation. https://b2bsol.panasonic.biz/semi-spt/apl/en/news/contents/2013/apec/panel/APEC_TMOS_FPD_3P.pdf

60. *EPC2050 – Enhancement-Mode Power Transistor – Preliminary Specification Sheet* (2018, April). Efficient Power Conversion Corporation.

61. *PGA26E07BA Preliminary Datasheet* (2016, October). Panasonic Corporation.

62. Sayadi, L., Iannaccone, G., Sicre, S., Häberlen, O., & Curatola, G. (2018). Threshold voltage instability in p-GaN gate AlGaN/GaN HFETs. *IEEE Transactions on Electron Devices, 65*(6), 2454–2460. ISSN: 0018-9383. https://doi.org/10.1109/TED.2018.2828702

63. Chellappan, S. (2017, November). *Design Considerations of GaN Devices for Improving Power-Converter Efficiency and Density*. Texas Instruments Incorporated.

64. Zuk, P. (n.d.). *Leading the GaN Revolution – Transphorm's GaN FET Technology Performance*. Transphorm. https://www.transphormusa.com/en/gan-training/.

65. Brohlin, P., Ramadass, Y., & Kaya, C. (2018, November). *Direct-Drive Configuration for GaN Devices*. Texas Instruments Incorporated.

66. Norling, K., Lindholm, C., & Draxelmayr, D. (2012, December). An optimized driver for SiC JFET-based switches enabling converter operation with more than 99% efficiency. *IEEE Journal of Solid-State Circuits, 47*(12), 3095-3104. ISSN: 0018-9200. https://doi.org/10.1109/JSSC.2012.2225736

67. Lidow, A. (2014). *GaN Transistors for Efficient Power Conversion*. Power Conversion Publications. ISBN: 978-1-118-84476-2.

68. Guz, M., Sanderlin, D., How Sin, B., de Rooij, M., McDonald, T., Le, P., et al. (2018, June). IEEE ITRW working group position paper – System integration and application: Gallium nitride. *IEEE Power Electronics Magazine, 5*(2), 34–39. FOM GaN Si. https://doi.org/10.1109/MPEL.2018.2821786

69. Chu, R., Hughes, B., Chen, M., Brown, D., Li, R., Khalil, S., et al. (2013). Normally-off GaN-on-si transistors enabling nanosecond power switching at one kilowatt. In: *71st Device Research Conference*.

70. Moench, S., Hillenbrand, P., Hengel, P., & Kallfass, I. (2017, October). Pulsed measurement of sub-nanosecond 1000 V/ns switching 600 V GaN HEMTs using 1.5 GHz low-impedance voltage probe and 50 ohm scope. In: *Proceedings of IEEE 5th Workshop Wide Bandgap Power Devices and Applications (WiPDA)* (pp. 132–137). https://doi.org/10.1109/WiPDA.2017.8170535

71. Bortis, D., Knecht, O., Neumayr, D., & Kolar, J. W. (May 2016). Comprehensive evaluation of GaN GIT in lowand high-frequency bridge leg applications. In: *2016 IEEE 8th International Power Electronics and Motion Control Conference (IPEMC-ECCE Asia)* (pp. 21–30). https://doi.org/10.1109/IPEMC.2016.7512256

72. Ebli, M., Wattenberg, M., & Pfost, M. (2017, December). A gate driver approach enabling switching loss reduction for hard-switching applications. In: *Proceedings of IEEE 12th International Conference Power Electronics and Drive Systems (PEDS)* (pp. 968–971). https://doi.org/10.1109/PEDS.2017.8289133

73. *Impacts of the dv/dt Rate on MOSFETs* (2018, September). Nexperia GmbH.

74. Persson, E. (2015). *Practical Application of 600 V GaN HEMTs in Power Electronics*. Infineon Technologies AG. https://www.infineon.com/dgdl/Infineon-Presentation_GalliumNitride_GaNApplications_SeminarAPEC2015-AP-v01_00-EN.pdf?fileId=5546d4624bcaebcf014c2c265e69007e

75. Huang, X., Lee, D. Y., Bondarenko, V., Baker, A., Sheridan, D. C., Huang, A. Q., et al. (2014, June). Experimental study of 650V AlGaN/GaN HEMT short-circuit safe operating area (SCSOA). In: *Proceedings of IEEE 26th International Symposium on Power Semiconductor Devices IC's (ISPSD)* (pp. 273–276). https://doi.org/10.1109/ISPSD.2014.6856029

76. Oeder, T., Castellazzi, A., & Pfost, M. (2017, May). Experimental study of the short-circuit performance for a 600V normally-off P-gate GaN HEMT. In: *Proceedings of 29th International Symposium on Power Semiconductor Devices and IC's (ISPSD)* (pp. 211–214). https://doi.org/10.23919/ISPSD.2017.7988925

77. Wu, H., Fayyaz, A., & Castellazzi, A. (2018, May). P-gate GaN HEMT gate-driver design for joint optimization of switching performance, freewheeling conduction and short-circuit robustness. In: *Proceedings of IEEE 30th International Symposium on Power Semiconductor Devices and ICs (ISPSD)* (pp. 232–235). https://doi.org/10.1109/ISPSD.2018.8393645

78. Castellazzi, A., Fayyaz, A., Zhu, S., Oeder, T., & Pfost, M. (2018, March). Single pulse short-circuit robustness and repetitive stress aging of GaN GITs. In: *Proceedings of IEEE*

International Reliability Physics Symposium (IRPS) (pp. 4E.1-1–4E.1-10). https://doi.org/10.1109/IRPS.2018.8353593

79. Huque, M. A., Islam, S. K., Tolbert, L. M., & Blalock, B. J. (2012). A 20°C universal gate driver integrated circuit for extreme environment applications. In: *IEEE Transactions on Power Electronics, 27*(9), 4153–4162. ISSN: 0885-8993. https://doi.org/10.1109/TPEL2012.2187934

80. März, M., Schletz, A., Eckardt, B., Egelkraut, S., & Rauh, H. (2010, March). Power electronics system integration for electric and hybrid vehicles. In: *Proceedings of 6th International Conference on Integrated Power Electronics Systems* (pp. 1–10).

81. Hensel, A., Wilhelm, C., & Kranzer, D. (2012, September). Application of a new 600 V GaN transistor in power electronics for PV systems. In: *2012 15th International Power Electronics and Motion Control Conference (EPE/PEMC)* (pp. DS3d.4-1–DS3d.4–5). https://doi.org/10.1109/EPEPEMC.2012.6397350

82. Ma, Y., Lin, Z., Lin, Y., Lee, C., Huang, T., Chen, K., et al. (2019, February). 29.6 A digital-type GaN driver with current-pulse-balancer technique achieving sub-nanosecond current pulse width for high-resolution and dynamic effective range LiDAR system. In: *Proceedings of IEEE International Solid- State Circuits Conference – (ISSCC)* (pp. 466–468)). https://doi.org/10.1109/ISSCC.2019.8662308

83. Butler, S. W. (2019). Enabling a powerful decade of changes. *IEEE Power Electronics Magazine, 6*(1), 18–20.

84. Corporation, Efficient Power Conversion (2019). *eGan® FETs and ICs for LiDAR – APPLICATION BRIEF: AB005.* Efficient Power Conversion Corporation. https://epc-co.com/epc/Portals/0/epc/documents/briefs/AB005%20eGaN%20FETs%20for%20LiDAR%20Applications.pdf

85. Wittmann, J., Funk, T., Rosahl, T., & Wicht, B. (2017, September). A 12–48 V wide-Vin 9–15 MHz soft-switching controlled resonant DCDC converter. In: *Proceedings of ESSCIRC 2017 – 43rd IEEE European Solid State Circuits Conference* (pp. 348–351). https://doi.org/10.1109/ESSCIRC.2017.8094597

86. Liu, Z., Lee, F. C., Li, Q., & Yang, Y. (2015, September). Design of GaN-based MHz totem-pole PFC rectifier. In: *2015 IEEE Energy Conversion Congress and Exposition (ECCE)* (pp. 682–688). https://doi.org/10.1109/ECCE.2015.7309755

87. Zhou, L., Wu, Y., Honea, J., & Wang, Z. (2015, May). High-efficiency true bridgeless totem pole PFC based on GaN HEMT: Design challenges and cost-effective solution. In: *Proceedings of PCIM Europe 2015; International Exhibition and Conference for Power Electronics, Intelligent Motion, Renewable Energy and Energy Management* (pp. 1–8).

88. Huber, L., Jang, Y., & Jovanovic, M. M. (2008). Performance evaluation of bridgeless PFC boost rectifiers. *IEEE Transactions on Power Electronics, 23*(3), 1381–1390. ISSN: 0885-8993. https://doi.org/10.1109/TPEL.2008.921107

89. Incorporation, GaN Systems (2017). *Gallium Nitride Power Transistors in the EV World.* GaN Systems Incorporation. https://gansystems.com/wp-content/uploads/2018/01/The-Benefits-of-Gallium-Nitride-Power-Transistors-Span-Multiple-Markets.pdf

90. *EPC2040 – Enhancement – Mode Power Transistor* (2018, June). Efficient Power Conversion Corporation.

91. *GS-065-120-1-D 650V – Enhancement Mode GaN Transistor* (2018). GaN Systems Incorporation. https://gansystems.com/gan-transistors/gs-065-120-1-d/

92. Lidow, A. (2012). *GaN Transistors for Efficient Power Conversion.* Efficient Power Conversion Corporation. https://epc-co.com/epc/Portals/0/epc/documents/presentations/IIC2012.pdf

93. *EPC2038 – Enhancement Mode Power Transistor With Integrated Reverse Gate Clamp Diode* (2018, October). Efficient Power Conversion Corporation.

94. Villamor, A., Dogmus, E., & Lin, H. (2018, November). *Power GaN 2018: Epitaxy, Devices, Applications and Technology Trends.* Yole Développment. https://www.i-micronews.com/images/Flyers/Power/YD18049_Power_GaN_2018_November_2018_Flyer.pdf

95. Barbarini, E., & Le Troadec, V. (2016, March). *GaN on Si HEMT vs SJ MOSFET: Technology and Cost Comparison Will SJ MOSFETs still be Attractive Compared to GaN Devices?* Yole Développment. https://www.i-micronews.com/images/Flyers/Power/Yole_GaN_on_Si_ HEMT_vs_SJ_MOSFET_technology_and_cost_comparison_March_2016.pdf

96. *Final Data Sheet EiceDRIVER™1ED020I12-F2 Single IGBT Driver IC* (2017, September). Infineon Technologies AG.

97. Grbovic, P. J. & Arpilliere, M. (2009, September). IGBT Cross Conduction Phenomenon — Origin and Simple Protection Gate Driving Technique. In: *Proceedings of 13th European Conference on Power Electronics and Applications* (pp. 1–10).

98. Zhang, Z. L., Dong, Z., Hu, D. D., Zou, X. W., & Ren, X. (2017, July). Three-level gate drivers for eGaN HEMTs in resonant converters. *IEEE Transactions on Power Electronics, 32*(7), 5527–5538. ISSN: 0885-8993. https://doi.org/10.1109/TPEL.2016.2606443

99. Sørensen, C., Fogsgaard, M. L., Christiansen, M. N., Graungaard, M. K., Nørgaard, J. B., Uhrenfeldt, C., et al. (2015, June). Conduction, reverse conduction and switching characteristics of GaN E-HEMT. In: *2015 IEEE 6th International Symposium on Power Electronics for Distributed Generation Systems (PEDG)* (pp. 1–7). https://doi.org/10.1109/ PEDG.2015.7223051

100. Dong, Z., Zhang, Z., Ren, X., Ruan, X., & Liu, Y. F. (2015, March). A gate drive circuit with mid-level voltage for GaN transistors in a 7-MHz isolated resonant converter. In: *2015 IEEE Applied Power Electronics Conference and Exposition (APEC)* (pp. 731–736). https://doi.org/ 10.1109/APEC.2015.7104431

101. Prasobhu, P. K., Raveendran, V., Buticchi, G., & Liserre, M. (2018). Active thermal control of GaN-based DC/DC converter. *IEEE Transactions on Industry Applications, 54*(4), 3529– 3540. ISSN: 0093-9994. https://doi.org/10.1109/TIA.2018.2809543

102. Dalton, J. J. O., Wang, J., Dymond, H. C. P., Liu, D., Pamunuwa, D., Stark, B. H., et al. (2017, March). Shaping switching waveforms in a 650 V GaN FET bridge-leg using 6.7 GHz active gate drivers. In: *Proceedings of IEEE Applied Power Electronics Conference and Exposition (APEC)* (pp. 1983–1989). https://doi.org/10.1109/APEC.2017.7930970

103. Groeger, J., Schindler, A., Wicht, B., & Norling, K. (2017, March). Optimized dv/dt, di/dt sensing for a digitally controlled slope shaping gate driver. In: *Proceedings of IEEE Applied Power Electronics Conference and Exposition (APEC)* (pp. 3564–3569). https://doi.org/10. 1109/APEC.2017.7931209

104. Sun, B., Burgos, R., Zhang, X., & Boroyevich, D. (2016, September). Active dv/dt Control of 600V GaN transistors. In: *Proceedings of IEEE Energy Conversion Congress and Exposition (ECCE)* (pp. 1–8). https://doi.org/10.1109/ECCE.2016.7854818

105. Perez, A., Jorda, X., Godignon, P., Vellvehi, M., Galvez, J. L., Millan, J., et al. (2003, September). An IGBT gate driver integrated circuit with full-bridge output stage and short circuit protections. In: *Semiconductor Conference, 2003. CAS 2003. International* (Vol. 2, p. 248). https://doi.org/10.1109/SMICND.2003.1252427

106. Schweber, B. (2018). *Motor Gate-Drive Isolation: Go Optocoupler Transformer or Other.* https://www.mouser.de/applications/motor-gate-drive-isolation/?utm_medium=email&utm_ campaign=elq-18.1121-techapp-industrialmotorcontrol-emea-de&utm_source=eloqua& subid=87b8c6476ad843f1a66e43d0579f8310&utm_content=5479745

107. *The ISO72x Family of High-Speed Digital Isolators* (2006, January). Infineon Technologies AG.

108. Ishigaki, M., Fafard, S., Masson, D. P., Wilkins, M. M., Valdivia, C. E., & Hinzer, K. (2017, March). A new optically-isolated power converter for 12 V gate drive power supplies applied to high voltage and high speed switching devices. In: *Proceedings of IEEE Applied Power Electronics Conference and Exposition (APEC)* (pp. 2312–2316). https://doi.org/10.1109/ APEC.2017.7931022

109. Park, S., & Jahns, T. (2005). A self-boost charge pump topology for a gate drive high-side power supply. In: *IEEE Transactions on Power Electronics, 20*(2), 300–307. ISSN: 0885-8993. https://doi.org/10.1109/TPEL.2004.843013

110. Lin, R., & Lee, F. (1997, June). Single-power-supply-based transformerless IGBT/MOSFET gate driver with 100% high-side turn-on duty cycle operation performance using auxiliary bootstrapped charge pumper. In: *Proceedings of IEEE Power Electronics Specialists Conference (PESC)* (Vol. 2, pp. 1205–1209). https://doi.org/10.1109/PESC.1997.616904

111. Mitova, R., Crebier, J.-C., Aubard, L., & Schaeffer, C. (2005). Fully integrated gate drive supply around power switches. *IEEE Transactions on Power Electronics, 20*(3), 650–659. ISSN: 0885-8993. https://doi.org/10.1109/TPEL.2005.846541

112. Crebier, J., & Rouger, N. (2008). Loss free gate driver unipolar power supply for high side power transistors. *IEEE Transactions on Power Electronics, 23*(3), 1565–1573. ISSN: 0885-8993. https://doi.org/10.1109/TPEL.2008.921163

113. *Application Note AN-978 – HV Floating MOS-Gate Driver ICs* (2008, March). How to generate a negative gate bias, in *International Rectifier*.

114. Nagai, S., Negoro, N., Fukuda, T., Otsuka, N., Sakai, H., Ueda, T., et al. (2012, February). A DC-isolated gate drive IC with drive-by microwave technology for power switching devices. In: *2012 IEEE International Solid-State Circuits Conference* (pp. 404–406). https://doi.org/10.1109./.ISSCC.2012.6177066

115. Sarnago, H., Lucia, O., Mediano, A., & Burdio, J. M. (2014). Design and implementation of a high-efficiency multiple-output resonant converter for induction heating applications featuring wide bandgap devices. In: *IEEE Transactions on Power Electronics, 29*(5), 2539–2549. ISSN: 0885-8993. https://doi.org/10.1109/TPEL.2013.2278718

116. Seidel, A., Costa, M., Joos, J., & Wicht, B. (2015b, March). Isolated 100% PWM gate driver with auxiliary energy and bidirectional FM/AM signal transmission via single transformer. In: *Proceedings of Applied Power Electronics Conference and Exposition, Charlotte, U.S.A.* (pp. 2581–2584). Piscataway: IEEE. https://doi.org/10.1109/APEC.2015.7104715

117. Roschatt, P. M., McMahon, R. A., & Pickering, S. (2015, June). Investigation of dead-time behaviour in GaN DC-DC buck converter with a negative gate voltage. In: *Proceedings of 9th International Conference on Power Electronics and ECCE Asia (ICPE-ECCE Asia)* (pp. 1047–1052). https://doi.org/10.1109/ICPE.2015.7167910

118. Rindfleisch, C., & Wicht, B. (2020, February). 11.3 A one-step 325V to 3.3-to-10V 0.5W resonant DC-DC converter with fully integrated power stage and 80.7% efficiency. In: *Proceedings of IEEE International Solid- State Circuits Conference – (ISSCC)* (pp. 194–196). https://doi.org/10.1109/ISSCC19947.2020.9063150

119. Ke, X., Sankman, J., Song, M. K., Forghani, P., & Ma, D. B. (2016, January). 16.8 A 3-to-40V 10-to-30MHz automotive-use GaN driver with active BST balancing and VSW dual-edge dead-time modulation achieving 8.3% efficiency improvement and 3.4ns constant propagation delay. In: *Proceedings of IEEE International Solid-State Circuits Conference (ISSCC)* (pp. 302–304). https://doi.org/10.1109/ISSCC.2016.7418027

120. Roschatt, P. M., Pickering, S., & McMahon, R. A. (2016). Bootstrap voltage and dead time behavior in GaN DC –DC buck converter with a negative gate voltage. *IEEE Transactions on Power Electronics, 31*(10), 7161–7170. ISSN: 0885-8993. https://doi.org/10.1109/TPEL.2015.2507860

121. *LMG1210 200-V, 1.5-A, 3-A Half-Bridge GaN Driver With Adjustable Dead Time* (2018, May). Texas Instruments Incorporated.

122. *4-A and 6-A High-Speed 5-V Drive, Optimized Single-Gate Driver* (2012, December). Texas Instruments Incorporated.

123. Chen, B. (2008, June). Isolated half-bridge gate driver with integrated high-side supply. In: *Power Electronics Specialists Conference, 2008. PESC 2008. IEEE* (pp. 3615–3618). https://doi.org/10.1109/PESC.2008.4592516

124. Kaeriyama, S., Uchida, S., Furumiya, M., Okada, M., & Mizuno, M. (2010, June). A 2.5kV isolation 35kV/us CMR 250Mbps 0.13mA/Mbps digital isolator in standard CMOS with an on-chip small transformer. In: *Proceedings of Symposium on VLSI Circuits* (pp. 197–198). https://doi.org/10.1109/VLSIC.2010.5560301

125. Ma, S., Zhao, T., & Chen, B. (2014, March). 4A isolated half-bridge gate driver with 4.5V to 18V output drive voltage. In: *2014 Twenty-Ninth Annual IEEE Applied Power Electronics Conference and Exposition (APEC)* (pp. 1490–1493). https://doi.org/10.1109/APEC.2014. 6803504

126. Colin, D., & Rouger, N. (2016, September). High speed optical gate driver for wide band gap power transistors. In: *Proceedings of IEEE Energy Conversion Congress and Exposition (ECCE)* (pp. 1–6). https://doi.org/10.1109/ECCE.2016.7855156

127. Rouger, N., Colin, D., Le, L. T., & Crébier, J. C. (2016, November). CMOS gate drivers with integrated optical interfaces for extremely fast power transistors. In: *Proceedings of Ship Propulsion and Road Vehicles Int 2016 International Conference on Electrical Systems for Aircraft, Railway Transportation Electrification Conference (ESARS-ITEC)* (pp. 1–6). https:// doi.org/10.1109/ESARS-ITEC.2016.7841366

128. *Si827x Data Sheet* (2018, May). Silicon Laboratories Inc.

129. Hackel, J., Seidel, A., Wittmann, J., & Wicht, B. (2016, September). Capacitive gate drive signal transmission with transient immunity up to 300 V/ns. In: *Proceedings of ANALOG 2016; 15. ITG/GMM-Symp* (pp. 1–5).

130. Zeltner, S. (2010, March). Insulating IGBT driver with PCB integrated capacitive coupling elements. In: *Proceedings of 6th International Conference on Integrated Power Electronics Systems* (pp. 1–6).

131. Vasic, D., Costa, F., & Sarraute, E. (2006). Piezoelectric transformer for integrated MOSFET and IGBT gate driver. *IEEE Transactions on Power Electronics, 21*(1), 56–65. ISSN: 0885-8993. https://doi.org/10.1109/TPEL.2005.861121

132. *L6384E – High-Voltage Half Bridge Driver* (2007, October). STMicroelectronics.

133. Txapartegi, M. G., & Gueguen, P. (2018). *Gate Drivers Market Evolution: Coreless Isolation and WBG Specific Solutions*. Yole Développment. https://www.psma.com/sites/default/files/ uploads/tech-forums-safety-compliance/presentations/is042-gate-drivers-market-evolution-coreless-isolation-and-wbg-specific-solutions.pdf

134. Nagai, S., Kawai, Y., Tabata, O., Fujiwara, H., Negoro, N., Ishida, M., et al. (2014, August). A drive-by-microwave isolated gate driver with gate current charge for IGBTs. In: *Proceedings of 16th European Conference on Power Electronics and Applications* (pp. 1–6). https://doi. org/10.1109/EPE.2014.6910757

135. *PARASITIC Inductance of Multilayer Ceramic CapacitorS* (n.d.). AVX Corporation.

136. *Quad Flatpack No-Lead Logic Packages* (2004, February). Texas Instruments Incorporated.

137. *AN-1205 Electrical Performance of Packages* (2004, May). Texas Instruments Incorporated.

138. AG, Infineon Technologies (2014). *Section 5 – High Speed PCB Layout Techniques*. Texas Instruments. http://www.ti.com/lit/ml/slyp173/slyp173.pdf

139. Stückler, F., Abdel-Rahman, S., & Siu, K. (2015, May). *600 V CoolMOS^{TM} C7 Design Guide*, Infineon Technologies AG.

Chapter 3
Gate Drivers Based on High-Voltage Charge Storing (HVCS)

This chapter presents a driver comprising an area-efficient gate driver output stage, which is enabled by the concept of high-voltage charge storing (HVCS), proposed as part of this work [1].[1] The gate driver output stage delivers a peak gate current of 2.6 A at a maximum driver output voltage $V_{GS,max} = 15$ V. Three options of the gate driver output stage based on HVCS [1] are described in Sect. 3.2. The gate driver circuit implementation is presented in Sect. 3.3. To maintain a permanently turned-on gate driver, various charge pump concepts are compared. A series regulator is proposed as a gate driver and bootstrap supply. Section 3.5 derives a sizing guideline for the required bootstrap capacitors. Experimental results in Sect. 3.6 verify the HVCS concept and the calculations of Sect. 3.5.

3.1 The Concept of HVCS

The conventional bootstrap circuit, described in Sect. 2.1.2, suffers from a small charge allocation Q_{tot} with respect to the whole charge stored in the bootstrap capacitor C_{B1}. According to the gate-charge equation,

$$Q_{tot} = C_{B1} \cdot \Delta V_1. \tag{3.1}$$

This is due to the small voltage drop ΔV_1 at the bootstrap capacitor after it discharges at gate driver turn-on. Figure 3.1a shows a two-NMOS driver output stage supplied by a conventional bootstrap circuit (see Sect. 2.1.2). Figure 3.1b shows a first implementation option of HVCS. The key idea of HVCS is to

---

[1]©2015 IEEE. Reprinted, with permission, from Seidel, A. et al. "Area Efficient Integrated Gate Drivers Based on High-Voltage Charge Storing". In: IEEE Journal of Solid-State Circuits 50.7, pp. 1550–1559. ISSN: 0018-9200. DOI: https://doi.org/10.1109/JSSC.2015.2410797.

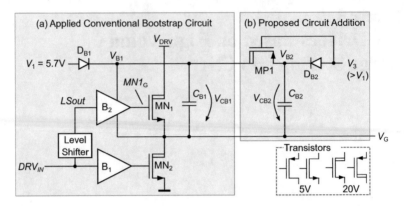

Fig. 3.1 Two NMOS transistor output stage buffer with (**a**) conventional bootstrap circuit and (**b**) extension by the proposed bootstrap circuit

initially charge a second high-voltage bootstrap capacitor C_{B2} to a higher voltage. Discharging C_{B2} to the low-voltage bootstrap capacitor voltage level V_{CB1}, a larger voltage drop occurs. For example, if it discharges from 15 V to 5 V, the large voltage swing of 10 V results in a large amount of charge (see (3.1)). In consequence, a significantly smaller bootstrap capacitor C_{B1} can be used. Area is saved even with the addition of capacitor C_{B2}. C_{B1} is finally connected in parallel to C_{B2}. It delivers a smaller amount of charge than C_{B2} and buffers V_{CB1} if C_{B2} delivers too much or little charge, preventing over- or undervoltage. In particular, this protects the gates from exceeding their maximum gate-source voltage. Section 3.2 gives a detailed description of the implementation option of the proposed bootstrap circuit in Fig. 3.1 and derives two other options.

3.2   Circuit Options of a Gate Driver Output Stage Based on HVCS

Figure 3.1 shows the first option of the bootstrap concept based on HVCS. The second bootstrap capacitor C_{B2} is charged to V_{CB2o} which is V_3 minus $V_{F,DB2}$ (forward voltage of D_{B2}), as shown in Fig. 3.2a with $V3 = 15$ V. For V_3, the same voltage as V_{DRV} could be used. C_{B1} is charged to $V_{CB1o} \sim 5$ V which is $V_1 - V_{F,DB1}$ (Fig. 3.2c). When $LSout$ changes to "high," the gate driver output voltage V_G is rising (Fig. 3.2b) and MP_1 begins to conduct as V_{B2} exceeds V_3 by more than $V_{th,MP1}$ of MP_1. C_{B2} discharges to C_{B1}, to the gate node of MN_1 and to the circuits supplied by V_{B1}. A charge balance between C_{B1}, C_{B2}, and any additional load capacitance (mainly the gate capacitance of MN_1) occurs. As a main advantage, the proposed bootstrap circuit can be easily added in parallel to a conventional bootstrap circuit.

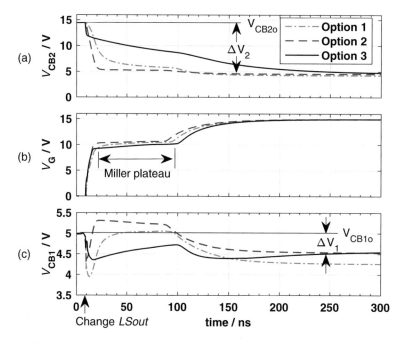

Fig. 3.2 Transient voltage signals of the three proposed bootstrap circuit options for $C_{B1} = 72.6\,\text{pF}$ and $C_{B2} = 18.4\,\text{pF}$

Due to a large voltage dip at C_{B2}, a considerable charge allocation can be realized as shown in (3.1). C_{B2} is charged up to 14.3 V ($= V_3 - V_F = 15\,\text{V} - 0.7\,\text{V}$), for example, and is discharged by $\Delta V_2 = 9.8\,\text{V}$ to 4.5 V. 4.5 V equals the voltage across C_{B1} after a voltage dip of $\Delta V_1 = 0.5\,\text{V}$ as depicted in Fig. 3.2c. In this example, the voltage dip $\Delta V_2 = 9.8\,\text{V} \sim 10\,\text{V}$ at V_{CB2} is nearly 20 times larger than the voltage dip of 0.5 V, if a comparable conventional bootstrap capacitor was used. In consequence, the proposed bootstrap capacitor can be about 20 times smaller to provide the same charge.

The circuit proposed in Fig. 3.1 has the drawback that MP_1 turns on after V_G rises. However, most of the charge is needed before. This causes a short voltage dip at V_{B1} at the beginning of a switching phase until C_{B1} is recharged from C_{B2} as shown in Fig. 3.2c for circuit option one (at $t \sim 10\,\text{ns}$). A voltage dip larger than specified can influence circuit blocks supplied from V_{B1} and may, for example, lead to faulty switching in the level shifter. In addition to this effect, the circuit of Fig. 3.1b requires a high-voltage PMOS transistor MP_1, occupying larger die area compared to its NMOS counterpart.

A second circuit option, which is presented in more detail in [1], has an additional C_{B2}-discharge path in parallel to MP_1 that conducts the current from the beginning of the switching phase. Since the second C_{B2}-discharge path delivers

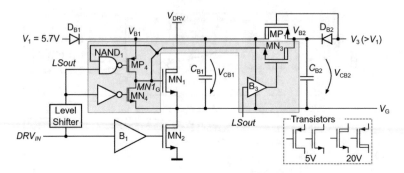

Fig. 3.3 Proposed bootstrap circuit with an NMOS transistor connecting directly to the gate of MN_1

charge independent of the gate charge demand of MN_1, an overshoot of C_{B1} may occur. This is demonstrated in Fig. 3.2c reaching almost the C_{B1} maximum ratings.

Figure 3.3 shows a third circuit option, which addresses these drawbacks. MN_3 directly connects C_{B2} to the gate node of MN_1 via the buffer B_3 when $LSout$="high", while MP_4 and MN_4 are in off-state. The charge for the gate node $MN_{1,G}$ is supplied directly from V_3 without discharging C_{B2} and C_{B1} before the driver output voltage V_G rises. This increases the voltage stability of V_{B1} and reduces the total required charge from C_{B2} and C_{B1} resulting in a lower voltage dip ΔV_1 (Fig. 3.2c). When $MN_{1,G}$ reaches the turn-off level of the logic gate $NAND_1$, MP_4 turns on, and $MN_{1,G}$ is fully charged by C_{B1}. During voltage rise at node $MN_{1,G}$, the gate-source voltage of MN_3 decreases and is finally turned-off. MP_1 still connects C_{B2} to C_{B1}.

There are several advantages of this circuit. The charge from V_3 and C_{B2} is immediately available during the whole switching phase. V_{B1} is more stable because the current peak caused by charging $MN_{1,G}$ in the beginning is provided by C_{B2} and has little effect on V_{B1} as Fig. 3.2c shows. Furthermore, the high-voltage PMOS transistor MP_1 is optional, because the low-voltage transistor MP_4 keeps MN_1 in its on-state. After driver turn-on, C_{B2} gets disconnected from C_{B1} if MP_1 is not implemented. This is an advantage, as the V_{CB1} voltage rail is automatically protected against excessive charge from C_{B2}, if it is initially charged to a higher voltage. Hence, it tolerates a wide input voltage range of V_3.

A fourth circuit implementation option, described in Sect. 4.2.5 (Fig. 4.16) [2], also tolerates a wide V_3 input-voltage range.

3.3 Driver Implementation

3.3.1 Charge Pump Concept

A two-NMOS driver output stage with a bootstrap circuit is limited in its on-state duration, because leakage currents discharge the bootstrap capacitors slowly. Hence, they cannot provide the gate overdrive of the pull-up NMOS transistor infinitely long. To enable a permanently turned-on output stage, the bootstrap capacitors must be recharged continuously. This is commonly done by a charge pump circuit [3, 4]. As a benefit, the proposed circuit requires only a small charge pump, because it has no constant supply currents.

In Fig. 3.4, different supply configurations for a charge pump are shown. The aim is to pump the charge to a voltage level $V_3 + V_1 = 20$ V when the output stage is in the on-state. For simplicity, the explanations below assume the diodes to be ideal with $V_F = 0$ V. In Fig. 3.4a, $V_3 = 15$ V is connected to the charge pump input and the inverter supply is $V_1 = 5$ V. When the oscillator output is "high," the inverter output is shorted to ground and the charge pump capacitor C_P is charged to the voltage level V_3 through diode D_{C1}. When the oscillator output turns "low," the inverter output and thus V_{CP} rises by $V1 = 5$ V and C_{B1} is charged via D_{C2} gradually towards $V_1 + V_3 = 20$ V. In this charge pump supply configuration, V_{B1} needs to be larger than V_3. Otherwise, a short circuit occurs between V_3 and V_{B1}, which may violate the maximum ratings at V_{B1}. This circuit is conceptual, showing that it is not suitable for the proposed gate driver output stage.

This is overcome in Fig. 3.4b with $V_1 = 5$ V as input voltage and $V_3 = 15$ V as inverter supply. Here, V_{B1} never falls below the input voltage V_1. For this option, an inverter INV_1 with high-voltage transistors is required. Furthermore, the pull-up transistor of the inverter needs to be controlled by a level shifter. A solution presented in [3] avoids level shifters, but suffers from losing three diode forward voltages in the power path of the charge pump.

Figure 3.4c shows the charge pump configuration, presented in [1], which is well suitable using the proposed bootstrap circuit. It is similar to the circuit in Fig. 3.4a, but with a low-voltage inverter. In contrast, it pumps the charge to V_{B2} instead of V_{B1}. If the driver is in the off-state, V_{B2} has the potential $V_3 = 15$ V. In the on-state, it rises up to $V_1 + V_3 = 20$ V. Thus, the voltage V_{B2} never falls below V_3 and no short current occurs. During driver on-state, the charge pump is activated. V_{B2} and thus C_{B1} and the connected circuit are supplied. This charge pump is implemented in the third option of the proposed bootstrap circuit, as part of the experimental gate driver IC that is described in Sect. 3.6. However, with this charge pump configuration, transistor MP_1 in the proposed bootstrap circuit (Fig. 3.3) is not optional anymore because it is required to connect V_{B1} with V_{B2}.

Figure 3.4d shows one more solution for the charge pump. It comprises a low-voltage inverter as well. Instead of V_3 or V_1, the floating ground V_G supplies the charge pump capacitor C_P. As V_G is always 5 V below V_{B1}, no crosscurrent occurs like in Fig. 3.4a. Applying this charge pump circuit to the third option of the

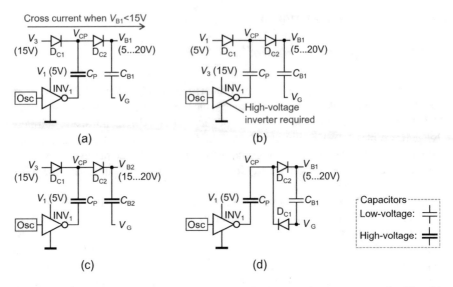

Fig. 3.4 Charge pump circuit with (**a**) 5 V inverter, 15 V input, and output connected to V_{B1}; (**b**) 15 V high-voltage inverter, 5 V input, and output connected to V_{B1}; (**c**) 5 V inverter, 15 V input, and output connected to V_{B2}; and (**d**) 5 V inverter, V_G input, and output connected to V_{B1}

proposed bootstrap circuit, the transistor MP_1 (Fig. 3.3) is not required anymore, because it pumps the charge directly to V_{B1} in contrast to the circuit in Fig. 3.4c.

Finally, the charge pump option in Fig. 3.4d is most suitable. If MP_1 is implemented anyway, option (c) is recommended as well.

3.3.2   Series Regulator

Figure 3.5 shows the linear regulator (LDO) [1], which is used to provide a stable and precise 5 V supply for the gate driver circuit. Due to D_{B1} (Fig. 3.3), one diode forward voltage is lost in the supply path of the bootstrap circuit. With $V_1 = 5$ V, this would result in a lower bootstrap capacitor voltage V_{CB1}. Yu and Arasu [5] replaces the diode D_{B1} by a synchronously switched high-voltage NMOS transistor. However, it needs additional circuits like a charge pump and level shifter. The LDO in Fig. 3.5 provides a less complex solution. It generates an output voltage V_{DD5V7}, which is one diode forward voltage above V_{DD5} (i.e., $V_{DD5V7} \sim 5.7$ V). This is achieved by the diode D_1, which has the same temperature and process dependencies like the diodes used in the bootstrap circuit. The LDO operates from a pre-regulated input voltage $V_{DD8} = 8$ V [6]. This approach is beneficial if such a higher voltage level is available on chip.

The LDO requires for start-up a minimum voltage-level at the supply rail V_{DD5}. Hence, a series regulator as part of the start-up circuit is implemented. It supplies

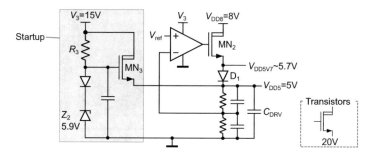

Fig. 3.5 Series regulator with two output voltages 5 V and 5.7 V

the V_{DD5} and V_{DD5V7} voltage domains from a stable supply $V_3 = 15$ V during start-up or in case of short current peaks at V_{DD5} or V_{DD5V7}. If the voltages at V_{DD5} or V_{DD5V7} are too low, the gate-source voltage of MN_3 is high and both supplies are charged from V_3 through MN_3. The series regulator is designed such that MN_3 with a threshold voltage of ~1 V shuts off once V_{DD5V7} and V_{DD5} reach their target voltage level.

V_{DD5V7} can also be used as a supply for the inverter of the charge pump, shown in Fig. 3.4c,d. In this case, a level shifter from 5 V to 5.7 V is required to control the inverter. If the maximum ratings of the low-voltage inverter are exceeded, high-voltage transistors need to be implemented. In this work, high-voltage transistors are used, because the available low-voltage transistors are rated for ≤ 5.5 V.

3.3.3 Level Shifters

3.3.3.1 Level Up Shifter

The level shifter circuit in the proposed bootstrap circuit like in Fig. 3.3 shifts the control signals to the high-side voltage domain in order to turn on or off the high-side transistor MN_1. Figure 3.6 shows the level-shifter circuit, which is an advanced and even faster option of the "ultra-fast mode" cross-coupled level shifter presented in [7]. The circuit is simplified for a better understanding of the basic function. The simplifications concern some delay compensating logic gates and the implementation of a power-on-reset circuit that are left out. Appendix 3.7 shows the complete circuit. Figure 3.7 shows the timing diagrams of the circuit in Fig. 3.6. For better understanding, the timings in Fig. 3.7a assume a constant floating ground potential V_G. Figure 3.7b considers a switching V_G node and is relevant for the description further down. At a turn-on signal ($LSin$="high"), MN_1 pulls down the node V_{xon} until the cascode transistor MP_1 turns off. MP_1 protects the low-voltage transistors connected to V_{xon} against overvoltage. The pull-down path for V_{xon} needs to be stronger than MP_3 in its on-state. The inverter I_1 detects a voltage dip at

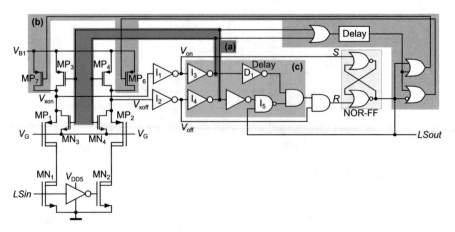

Fig. 3.6 Simplified level shifter circuit

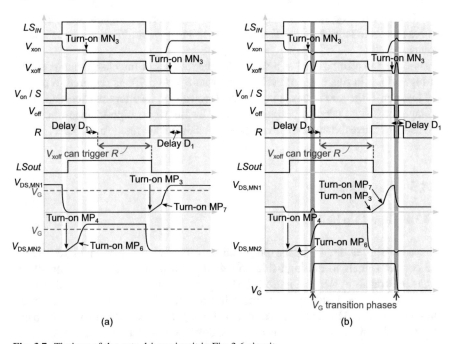

Fig. 3.7 Timings of the gate driver circuit in Fig. 3.6 circuit

V_{xon} turning V_{on} "high" in order to set the NOR flip-flop. For faster detection, the inverter I_1 is designed asymmetrically to be sensitive to small V_{xon} voltage dips. In conventional level shifter designs, a "high" signal at the flip-flop's set input does not propagate immediately to its output $LSout$ since its reset input is still "high." This is because the reset signal depends on the node V_{xoff}, which first needs to turn "high" via the inverter (MP_4, MN_4) charging the node capacitance at the drain of

MN$_2$. This event is triggered by the feedback path from the inverter I$_3$ (Fig. 3.6a). For a faster pull-up of node V_{xoff}, the strong transistor MP$_6$ turns-on for a short time, controlled by the logic in Fig. 3.6b which detects if both nodes V_{xoff} and V_{xon} are "high" [7]. The level shifter turn-off proceeds similarly to the level-shifter turn-on event pulling down V_{xoff} via MN$_2$ and MP$_2$ and setting the reset pulse of the flip-flop. A "high" signal at the reset input of a NOR flip-flop immediately propagates to its output turning $LSout$ to "low" independently of the flip-flop set input. To be rapidly ready for the next level shifter turn-on, MP$_7$ is implemented for a fast reset of V_{xon} to "high," accordingly to MP$_6$ for a faster level-shifter turn-on.

The turn-on event of the level shifter is slowed down significantly if the flip-flop needs to wait for a "low" state at its reset input, until it propagates a set pulse to its output. Therefore, this design comprises a logic part, shown in Fig. 3.6c. It turns the reset input of the flip-flop to "low," already a few nanoseconds after a level shifter turn-off event, via I$_3$ and the delay element D$_1$. I$_5$ keeps the reset input of the flip-flop "low" during level shifter off-state ($LSout$="low") and level shifter turn-on, i.e., until V_{xoff} turned to "high." When a level-shifter turn-on event occurs, the reset input of the flip-flop is already "low" and the set pulse can immediately turn $LSout$ to "high."

The delay element D$_1$ of a few nanoseconds is important as during that time the circuit is robust against a rising transition of the floating ground potential V_{G}. Figure 3.7b shows the level shifter timings for a transitioning V_{G}. A rising V_{G} leads to a charging of the parasitic capacitances between the nodes $V_{\text{xon}}/V_{\text{xoff}}$ and the level-shifter low-side pulling down $V_{\text{xon}}/V_{\text{xoff}}$ (see Sect. 2.4.5). This results in a "high" signal at V_{off}. During the delay provided by D$_1$, the reset input of the flip-flop is "low" keeping its output $LSout$ safely in "high"-state. Otherwise an unintended turn-off of the level shifter can occur. The delay D$_1$ prevents also an unintended turn-on of the level shifter in case of a rising V_{G} transition shortly after a level-shifter turn-off. This is not critical in applications like gate drivers where the rising edge of V_{G} always happens shortly after a turn-on event. Lutz et al. [8] presents a solution that detects a V_{G} transition (in this case, the falling edge is critical) with a logic block that prevents the false signals to propagate to the level shifter output. The logic block is also utilized for a fast level-shifter turn-on by providing a "low" signal at the reset input of the NOR flip-flop before turning-on the level shifter, similar to the approach in this work.

3.3.3.2 Level Down Shifter

To avoid crosscurrents in the gate driver output stage, like in Fig. 3.3, during driver turn-off, the high-side NMOS transistor MN$_1$ first has to switch off before turning on the low-side transistor MN$_2$. This is ensured by a level down shifter, shown in Fig. 3.8, which propagates a driver turn-off signal (DRV_{IN}="low") only when MN$_1$ is in the off-state. The MN$_1$-off-state information ($LSout$ = "low") is shifted down by turning on the high-voltage PMOS transistor MP$_2$ that pulls up the node $MN1_{off}$ via the cascode transistor MN$_3$. As soon as $MN1_{off}$ is "high," the logic gate I$_2$ is

Fig. 3.8 Level down shifter circuit

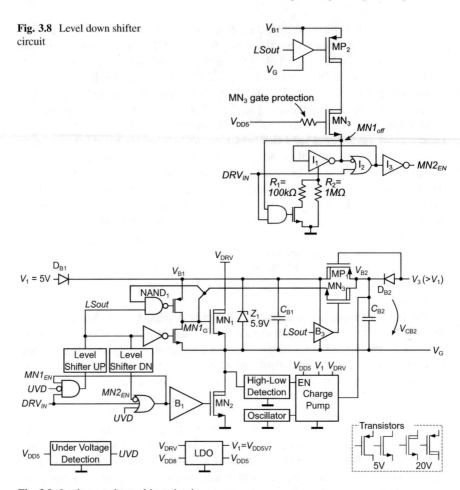

Fig. 3.9 Implemented gate driver circuit

sensitive to a DRV_{IN}="low" state, turning $MN2_{EN}$ "high" and enabling a safe driver output turn-off operation without crosscurrents. As the cascode MN_3 cannot fully pull up the node $MN1_{off}$, the inverter I_1 pulls $MN1_{off}$ fully to V_{DD5} via the feedback path from I_2. A driver turn-on event (DRV_{IN} = "high") immediately turns-off MN_2, and I_1 resets $MN1_{off}$ by pulling the node to ground via the resistors R_1 and R_2. R_1 enables a faster reset of $MN1_{off}$ and is only shortly activated until $MN1_{off}$ reaches "low" level. Now, the level down shifter is ready for the next driver turn-off event.

3.3.4   Gate Driver Circuit

The bootstrap circuit according to Fig. 3.3 turned out to be most advantageous and is therefore chosen for implementation. An overview of the whole gate driver circuit including the charge pump, oscillator, and the LDO is shown in Fig. 3.9. The additional zener diode Z_1 is inserted for failure protection against overvoltage in case that C_{B2} allocates too much charge. This could happen if node V_G falls one diode forward below ground due to ringing at the gate driver output, overcharging C_{B1} and C_{B2} (see also Sect. 2.4.2). The charge pump circuit is activated when the driver output stage is turned on, which is detected by the "high-low detection" circuit. The oscillator is a relaxation oscillator, running at a frequency of $\sim$2 MHz [6]. The level down shifter indicates that the switching of the level up shifter was done correctly. Furthermore, it generates a dead time to ensure that MN_2 turns on not before MN_1 is turned off. The signal $MN1\_EN$ gets "high" when MN_2 is in off-state to enable MN_1 to be turned on. The dead time is about 5 ns, according to a transistor-level simulation. This avoids a crosscurrent in the output stage. The undervoltage detection circuit monitors the voltage V_{DD5} and causes to pull-down the driver output node if an undervoltage is detected.

3.4   Further Applications of the Proposed Bootstrap Circuit

The proposed bootstrap circuit can not only be applied to output stages in gate driver circuits, but also in other applications with fully integrated power stages, e.g., integrated switched mode power supplies [9] or class D output stages [10, 11]. The larger size of PMOS transistors (like in Fig. 2.3a) leads to a preferred use of two NMOS transistors in the output stage (Fig. 2.3b) [9, 10]. The required bootstrap capacitor is often implemented as an external component because of its size. An output stage configuration as shown in Fig. 2.3c is not suitable, because its output node supports a high-current capability only during rising edge. When the output stage is in on-state, the strong pull-up NMOS transistor turns off and only the weak PMOS MP_1 is active.

The concept of HVCS can be also applied to buffer capacitors delivering the whole charge for an external power transistor. The concept of HVCS may deliver sufficient charge for devices with small gate charges. This way, HVCS is well suitable in particular to small or medium gallium nitride (GaN) transistors, which require significantly smaller gate charge. For example, a GaN transistor of type EPC8010 (EPC), rated for 100 V, 4 A, requires a maximum gate charge of $Q_G = 480$ pC and a gate-source voltage of $VGS = 5$ V. Using a conventional bootstrap circuit with a capacitor charged to 5.5 V, a voltage dip of 0.5 V results in a bootstrap capacitor of 960 pF, according to (3.1). Such a capacitance is too large to be integrated in standard CMOS technologies. With the proposed bootstrap circuit and HVCS, the second bootstrap capacitor C_{B2} can be discharged from 20 V to 5 V, resulting in a large voltage dip of 15 V. Assuming that the whole charge is allocated by C_{B2}, a capacitor value of only 32 pF would be required, Eq. (3.1). This represents a capacitance reduction by a factor of 30, down to a value, that could be integrated on-chip even with a high-voltage capacitor type. Sizing guidelines with exact calculations for an area-efficient sizing of the bootstrap capacitors including corner parameters are presented in Sect. 3.5.

3.5   Bootstrap Capacitor Sizing

3.5.1   Sizing Equations

The design goals below are defined for the sizing of C_{B1} and C_{B2} (Figs. 3.1 and 3.3).

1. Undervoltage: A maximum given voltage dip $\Delta V_{1,\max}$ must not be exceeded to ensure a proper circuit function in the worst case of operation and in process corner.
2. Overvoltage: The maximum overvoltage specifications need to be observed for all devices across all corners. A negative ΔV_1 describes the overvoltage at V_{B1}.
3. Area: The most area-efficient values for both capacitors should be obtained.

The following sizing guideline can be applied to all proposed circuit options (see Figs. 3.1 and 3.3, and [1]). C_{B2} provides a charge depending on the voltage dip ΔV_2 at C_{B2} in addition to the charge allocation of conventional bootstrapping (see (3.1)). ΔV_2 is defined as $V_{CB2o} - V_{CB1o} + \Delta V_1$. V_{CB1o} and V_{CB2o} are the voltages across C_{B1} and C_{B2} at the beginning of their discharge process. The total provided charge Q_{tot} can be calculated as

$$Q_{tot} = k_1 \cdot C_{B1} \cdot \Delta V_1 + k_2 \cdot C_{B2} \cdot \Delta V_2. \tag{3.2}$$

k_1 and k_2 are scaling factors of C_{B1} and C_{B2}, accounting for process tolerances. They are > 1 for a positive, < 1 for a negative capacitance tolerance. Equation (3.2) has to satisfy design goal (1) and (2) for worst-case conditions. Hence, a set of corner parameters for both design goals can be derived (Table 3.1). All values are available except for C_{B1} and C_{B2}. Inserting the parameter sets of Table 3.1 into (3.2) leads to two equations, which can be rearranged to yield C_{B1}:

$$C_{B1} = \frac{Q_{\max} - k_{2,min} \cdot C_{B2} \cdot \Delta V_{2,\min}}{k_{1,min} \cdot \Delta V_{1,\max}}, \tag{3.3}$$

$$C_{B1} = \frac{Q_{\min} - k_{2,max} \cdot C_{B2} \cdot \Delta V_{2,\max}}{k_{1,min} \cdot \Delta V_{1,\min}}. \tag{3.4}$$

$\Delta V_{2,\min}$ and $\Delta V_{2,\max}$ are defined as

Table 3.1 Two sets of corner parameters for design goal (1) and (2)

	ΔV_1	Q_{tot}	V_{CB1o}	V_{CB2o}	k_1	k_2
Goal (1)	max	max	min	min	min	min
Goal (2)	min	min	max	max	min	max

$$\Delta V_{2,min} = V_{CB2o,min} - V_{CB1o,min} + \Delta V_{1,max}, \tag{3.5}$$

and

$$\Delta V_{2,max} = V_{CB2o,max} - V_{CB1o,max} + \Delta V_{1,min}, \tag{3.6}$$

respectively. This way, (3.3) and (3.4) result in

$$C_{B2} = \frac{Q_{max} \cdot \Delta V_{1,min} - Q_{min} \cdot \Delta V_{1,max}}{\Delta V_{2,min} \cdot \Delta V_{1,min} \cdot k_{2,min} - k_{2,max} \cdot \Delta V_{2,max} \cdot \Delta V_{1,max}} \tag{3.7}$$

The value for C_{B2} can be inserted into (3.3) or (3.4) to calculate C_{B1}. This calculation approach results automatically in the smallest total layout area according to design goal (3). In case of stacked capacitors C_{B1} and C_{B2}, the same area for both components may lead to an optimal dimensioning, described below in Sect. 3.5.2.

3.5.2   Sizing Equations for Stacked Bootstrap Capacitors

In some technologies, the low voltage capacitance C_{B1} and high-voltage capacitance C_{B2} are implemented in different layers and have different capacitance densities. For example, a poly-nwell capacitor can be utilized for capacitor C_{B1} and a metal-metal capacitor with higher voltage rating for C_{B2}. In that case, C_{B1} and C_{B2} can be placed on top of each other. Choosing the same area for both capacitors may result in the most area-efficient layout. Hence, the capacitance ratio is determined by the capacitance density ratios of C_{B2} and C_{B1}. To ensure design goals (1) and (2), this ratio must not exceed the ratio between C_{B2} and C_{B1} calculated from the corners in Table 3.1. A higher ratio would cause an overshoot of V_{B1} (C_{B2} too large). A lower ratio is acceptable as it results in a higher value for C_{B1}, which buffers V_{B1} in positive and negative direction. Generally, the safety margin is enhanced by increasing the capacitance value of C_{B1}.

With a given nominal capacitance ratio $r = C_{B2}/C_{B1}$ and the worst-case parameters of goal (1) from Table 3.1, C_{B1} can be calculated from (3.2):

$$C_{B1} = \frac{Q_{tot}}{(k_1 \cdot \Delta V_1 + r \cdot k_2 \cdot \Delta V_2)}. \tag{3.8}$$

In a second step, C_{B2} is determined from $r = C_{B2}/C_{B1}$ with C_{B1} from (3.8).

3.5.3   Sizing Example

The parameters used in the transient simulations in Fig. 3.2 serve as an example for the sizing of C_{B1} and C_{B2}. Based on a transistor-level simulation, the total required charge is $Q_{tot} = 216\,\mathrm{pC}$ with an assumed deviation of $\pm 30\,\mathrm{pC}$. This assumption can be derived from the capacitance variation of a poly-nwell capacitance (5 %), as the gate capacitance of MN_1 (see Figs. 3.1 and 3.3) mainly determines Q_{tot}. For safety, a three times larger deviation of 15 % was chosen. The main charge is required for driving MN_1, which is a high-voltage transistor with $R_{DS,on} \sim 1\,\Omega$. The initial bootstrap capacitor voltages $V_{CB1o} = 5\pm 0.2\,\mathrm{V}$ and $V_{CB2o} = 14.3\pm 0.3\,\mathrm{V}$ are defined based on the tolerances of V_1 and V_3. $\Delta V_{1,min} = -0.5\,\mathrm{V}$, $\Delta V_{1,max} = 1\,\mathrm{V}$ are values, chosen according to design goal (1) and (2), respectively. The capacitance tolerances of ± 5 % of C_{B1} and ± 10 % of C_{B2} are technology-depending values, resulting in $k_{1,min} = 1 - 0.05 = 0.95$, $k_{1,max} = 1 + 0.05 = 1.05$, $k_{2,min} = 1 - 0.1 = 0.9$, and $k_{2,max} = 1 + 0.1 = 1.1$. C_{B2} results in 21.3 pF according to (3.3) or (3.4). With the given parameters and (3.7), C_{B1} results in 51.3 pF. The ratio must be smaller than $r = C_{B2}/C_{B1} = 0.39$.

In case that the technology allows stacking C_{B2} on top of C_{B1} (see Sect. 3.5.2), a technology-defined capacitance density ratio of $r = 0.25$ is assumed for a calculation example. This results in $C_{B1} = 75.8\,\mathrm{pF}$ (see (3.8)) and $C_{B2} = r \cdot C_{B1} = 19\,\mathrm{pF}$, both occupying the same geometrical area.

The simulations shown in Fig. 3.2 are based on similar values. A nominal voltage dip of $\Delta V_1 = 0.49\,\mathrm{V}$ is calculated from (3.2) using nominal values and $C_{B1} = 72.6\,\mathrm{pF}$ and $C_{B2} = 18.4\,\mathrm{pF}$. This value of ΔV_1 matches with the simulation of circuit options two and three in Fig. 3.2c. Circuit option one requires a higher Q_{tot} by 17 pC because V_3 do not provide charge at the beginning of the driver turn-on event in contrast to option two and three.

3.5.4   Comparison with Conventional Bootstrap Circuit

A comparison to a conventional bootstrap circuit can be based on (3.1). With $Q_{max} = 246\,\mathrm{pC}$ and $\Delta V_{1,max} = 1\,\mathrm{V}$, C_{B1} results in 246 pF. The example shows that the C_{B1} value and area reduces by about 70 % in case that $C_{B2} = 19\,\mathrm{pF}$ can be placed on top of $C_{B1} = 75.8\,\mathrm{pF}$, enabled by the proposed HVCS bootstrap circuit. Even if C_{B2} is placed next to C_{B1}, a decrease of about 38 % in area is achieved, assuming that C_{B2} has a capacitance density of 25 % of C_{B1}.

3.6 Experimental Results

The high-voltage gate driver circuit was manufactured in a 0.18 µm BCD technology (Fig. 3.10). The bootstrap circuit is designed with similar parameters as in the example in Sect. 3.5 and in the transient transistor-level simulation in Fig. 3.2a–c. The size of the output stage (MN_1 and MN_2) is about 0.25 mm x 0.45 mm. A metal-metal capacitor C_{B2} is on top of a poly-nwell capacitor C_{B1} with a size of 0.17 mm x 0.38 mm. A 600 V superjunction MOSFET (CoolMOS^{TM} generation C6, see Sect. 2.2.1) is connected at the driver output. The correct gate driver operation is tested and confirmed in a 200 V application, switching the high-side power transistor, shown in Fig. 3.11.

Measurements of the gate driver power consumption were performed with a load capacitor of 10 nF (typical gate capacitance of a Si power transistor) connected at the driver output at a switching frequency of 100 kHz and a driver output voltage of 15 V. Due to a much larger voltage swing at the bootstrap capacitor, the losses in the proposed gate driver circuit are higher referred to an ideal gate driver (see

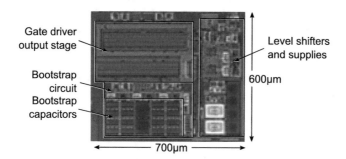

Fig. 3.10 Microphotograph of the proposed bootstrap circuit and gate driver

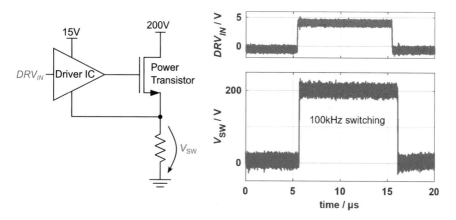

Fig. 3.11 Gate driver IC tested in an application

Sect. 4.1.3). The measured power consumption is 227 mW. According to Sect. 2.1.7, an ideal non-resonant gate driver would consume 225 mW ($10\,\text{nF} \cdot (15\,\text{V})^2 \cdot 100\,\text{kHz}$) for the charging and discharging of the load capacitor. Hence, the increase in losses is small. The main reason is that the HVCS concept is only applied to deliver relatively small charges compared to the gate charge of the power transistor.

A probe with a small input capacitance of 100 fF is used to measure the voltage V_{B1} to keep the impact on the measurement at a minimum. The oscilloscope ground has been tied to the V_G node to not exceed the maximum rated input voltage of the probe of 8 V. Figure 3.11 shows the measured transients of the voltage V_{B1} at turn-on of the power transistor for $V_3 = 15$ V. The behavior matches well with the simulation of Fig. 3.2c. The measured internal node voltage V_{CB2o} is 14.42 V, one diode drop less than V_3. The curve of V_{B1} shows an undershoot of approximately 1.3 V and settles at a voltage dip of 0.96 V. This value meets the maximum specified and calculated voltage dip of 1 V.

According to (3.1) and (3.2), the amount of charge stored in C_{B2} scales with its voltage, i.e., with V_3. Therefore, V_{B1} was also measured for V_3 at 13 V and 14 V to show the benefit of high-voltage charge storing in the second bootstrap capacitor C_{B2} (Fig. 3.12). The voltage dip gets significantly larger with a lower V_3. If V_3 was reduced down to $V_1 \sim 5.7$ V, a comparable charge allocation to a conventional bootstrap circuit with the same capacitor area would be achieved. In the real circuit, already for $V_3 = 12.5$ V the voltage dip gets too large to maintain a proper switching of the driver components.

In order to validate the calculation guideline provided in Sect. 3.5, the voltage dip can be calculated by inserting $\Delta V_2 = V_{CB2o} - V_{CB1o} + \Delta V_1$ into (3.2). Rearranging yields

$$\Delta V_1 = \frac{Q_{\text{tot}} - C_{B2} \cdot (V_{CB2o} - V_{CB1o})}{C_{B1} + C_{B2}}. \tag{3.9}$$

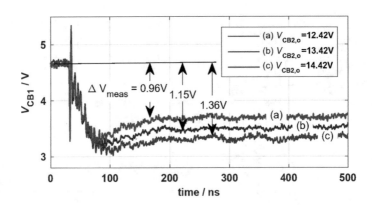

Fig. 3.12 Measurements of V_{B1} for various values of V_{CB2o}

Table 3.2 Comparison of measured ΔV_1 values with calculated values

	Calculated			Measured
V_{CB2o}	$\Delta V_{1,min}$	$\Delta V_{1,nom}$	$\Delta V_{1,max}$	ΔV_{meas}
12.3 V	0.34 V	0.82 V	1.38 V	1.36 V
13.3 V	0.11 V	0.62 V	1.19 V	1.15 V
14.3 V	0.41 V	−0.11 V	1.00 V	0.96 V

Table 3.3 Comparison of published gate driver circuits

Publication	This work	2014 [12]	2013 [13]	2007 [14]
Area (mm²)	0.42	0.17	8.5	1.16
Peak current I_{peak} (A)	2.6	1.5	5	2.6
Max. driver output voltage $V_{GS,max}$ (V)	15	15	30	30
FOM (A V mm⁻¹) ($I_{peak} \cdot V_{GS,max}/Area$)	93	132	18	67
Bootstrap capacitor	On-chip	Not required	On-chip	External
Technology	0.18 μm BCD	0.35 μm HV CMOS	SOI	0.8 μm BCD SOI
HV PMOS required	No	Yes	No	No

V_{CB1o} and V_{CB2o} (see Fig. 3.2) are available from measurements. Q_{tot} and the capacitor values were taken from simulation and are inserted as corner parameters according to goal (1) and (2). Table 3.2 confirms that the measured voltage dip values ΔV_{meas} (see Fig. 3.12) are all in the range of the maximum and minimum calculated values $\Delta V_{1,min}$ and $\Delta V_{1,max}$.

3.7 Comparison with Prior Art

For comparison, the characteristics of previously published gate driver designs are listed in Table 3.3. The indicated area, partly extracted from the die microphotograph, includes mainly the output stage, the level shifter, and supplies like the charge pump circuit. To show the area efficiency of a gate driver, a figure of merit (FOM) is introduced. Since the chip area of a gate driver output stage ($Area$) is approximately proportional to its current capability (I_{peak}) and to the maximum driver output voltage ($V_{GS,max}$), the figure of merit is proposed as $FOM = I_{peak} \cdot V_{GS,max}/Area$. The gate driver in [14] uses an external bootstrap capacitor resulting in a higher FOM than the driver in [13]. The driver in [15] is very compact, but relies on high-voltage PMOS transistors. Xu et al. [15] is not listed in Table 3.2 because of missing information about technology and chip layout. The circuit in [12] proposes a very compact driver output stage with a circuit configuration shown in Fig. 2.3a. In [12] the high-voltage PMOS transistors are less than two times larger in its specific resistance compared to the corresponding NMOS devices. This example shows that a conventional output stage configuration (Fig. 2.3a or c) might be more efficient in area, if appropriate high-voltage PMOS transistors are available. In this work, a very good FOM with a fully integrated output stage is achieved. Using the third option of the proposed bootstrap circuit (Fig. 3.3), no high-voltage PMOS transistor is required. In case of using the proposed charge pump circuit in Fig. 3.4c, only a small PMOS transistor is needed, enabling a permanently turned-on gate driver.

Appendix: Level Shifter Circuit

See Fig. 3.13.

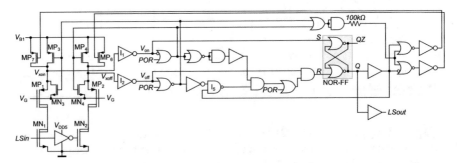

Fig. 3.13 Complete level shifter circuit in the gate driver based on HVCS, see Sect. 3.3.3 (Fig. 3.6)

References

1. Seidel, A., Costa, M. S., Joos, J., & Wicht, B. (2015). Area efficient integrated gate drivers based on high-voltage charge storing. *IEEE Journal of Solid-State Circuits, 50*(7), 1550–1559. ISSN: 0018-9200. https://doi.org/10.1109/JSSC.2015.2410797
2. Seidel, A., & Wicht, B. (2017, February). 25.3 A 1.3a gate driver for GaN with fully integrated gate charge buffer capacitor delivering 11nC enabled by high-voltage energy storing. In *2017 IEEE International Solid-State Circuits Conference (ISSCC)* (pp. 432–433). https://doi.org/10.1109/ISSCC.2017.7870446
3. Park, S., & Jahns, T. (2005). A self-boost charge pump topology for a gate drive high-side power supply. *IEEE Transactions on Power Electronics, 20*(2), 300–307. ISSN: 0885-8993. https://doi.org/10.1109/TPEL.2004.843013
4. Lin, R., & Lee, F. (1997, June). Single-power-supply-based transformerless IGBT/MOSFET gate driver with 100% high-side turn-on duty cycle operation performance using auxiliary bootstrapped charge pumper. In *Proceedings of IEEE Power Electronics Specialists Conference (PESC)* (Vol. 2, pp. 1205–1209). https://doi.org/10.1109/PESC.1997.616904
5. Yu, J., & Arasu, M. A. (2014, December). Transducer driver with active bootstrap switch. In *Proceedings of International Symposium on Integrated Circuits (ISIC)* (pp. 484–487). https://doi.org/10.1109/ISICIR.2014.7029480

6. Seidel, A., Costa, M., Joos, J., & Wicht, B. (2015, March). Isolated 100% PWM gate driver with auxiliary energy and bidirectional FM/AM signal transmission via single transformer. In *Proceedings of Applied Power Electronics Conference and Exposition, Charlotte, U.S.A.* (pp. 2581–2584). Piscataway: IEEE. https://doi.org/10.1109/APEC.2015.7104715
7. Moghe, Y., Lehmann, T., & Piessens, T. (2011, February). Nanosecond delay floating high voltage level shifters in a 0.35 μm HV-CMOS technology. *IEEE Journal of Solid-State Circuits, 46*(2), 485–497. ISSN: 0018-9200. https://doi.org/10.1109/JSSC.2010.2091322
8. Lutz, D., Seidel, A., & Wicht, B. (2018, September). A 50V, 1.45ns, 4.1pJ high-speed low-power level shifter for high-voltage DCDC converters. In *Proceedings of ESSCIRC 2018 – IEEE 44th European Solid State Circuits Conference (ESSCIRC)* (pp. 126–129). https://doi.org/10.1109/ESSCIRC.2018.8494292
9. Ng, J. C. W., Trescases, O., & Wei, G. (2007, December). Output stages for integrated DC-DC converters and power ICs. In *Proceedings of IEEE Conference on Electron Devices and Solid-State Circuits, 2007. EDSSC 2007* (pp. 91–94). https://doi.org/10.1109/EDSSC.2007.4450069
10. Berkhout, M. (2003, February). A class D output stage with zero dead time. In *Solid-State Circuits Conference, 2003. Digest of Technical Papers. ISSCC. 2003 IEEE International* (pp. 134–135). https://doi.org/10.1109/ISSCC.2003.1234237
11. Feng, Y., Wei, G., Ng, W., & Sugimoto, T. (2009, June). A 38W digital class D audio power amplifier output stage with integrated protection circuits. In *21st International Symposium on Power Semiconductor Devices IC's, 2009. ISPSD 2009* (pp. 53–56). https://doi.org/10.1109/ISPSD.2009.5157999
12. To, N.-D., Rouger, N., Arnould, J.-D., Corrao, N., Crebier, J.-C., & Lembeye, Y. (2014, February). Integrated gate driver circuits with an ultra-compact design and high level of galvanic isolation for power transistors. In *2014 8th International Conference on Integrated Power Systems (CIPS)* (pp. 1–6).
13. Greenwell, R. L., McCue, B. M., Tolbert, L. M., Blalock, B. J., & Islam, S. K. (2013, March). High-temperature SOI-based gate driver IC for WBG power switches. In *Applied Power Electronics Conference and Exposition (APEC), 2013 Twenty-Eighth Annual IEEE* (pp. 1768–1775). https://doi.org/10.1109/APEC.2013.6520535
14. Huque, M. A., Vijayaraghavan, R., Zhang, M., Blalock, B. J., Tolbert, L. M., & Islam, S. K. (2007, June). An SOI-based high-voltage, high-temperature gate-driver for SiC FET. In: *Power Electronics Specialists Conference, 2007. PESC 2007. IEEE* (pp. 1491–1495). https://doi.org/10.1109/PESC.2007.4342215
15. Xu., J., Sheng, L., & Dong, X. (2012, September). A novel high speed and high current FET driver with floating ground and integrated charge pump. In *Energy Conversion Congress and Exposition (ECCE), 2012 IEEE* (pp. 2604–2609). https://doi.org/10.1109/ECCE.2012.6342393

Chapter 4
Gate Drivers Based on High-Voltage Energy Storing (HVES)

The gate driver in Chap. 3 comprises a bootstrap capacitor, which could be fully integrated on-chip, due to the concept of high-voltage charge storing (HVCS). This capacitor buffers the supply rail of the pre-driver in the gate driver output stage. However, the buffer capacitor of the gate driver itself is not on-chip, as the gate driver is designed for silicon power transistors, which, typically, require a very large gate charge. Even with the concept of HVCS, a buffer capacitor integration is not suitable due to its large value. Furthermore, as will be described in Sect. 4.4.2, the energy dissipation of HVCS may not be negligible anymore for large values of gate charge.

Gallium nitride (GaN) transistors have $\sim 10\times$ smaller gate charge Q_G (see Sect. 2.2.2) compared to silicon devices. According to Sect. 3.4, the concept of HVCS may deliver sufficient charge for small or medium GaN transistors. To provide a solution for most of the currently available GaN and small silicon transistors, this chapter presents gate drivers based on a resonant capacitor implementation concept, called high-voltage energy storing (HVES), proposed as part of this work, first published in [1, 2].[1] Section 4.1 discusses the theory of HVES and design aspects in comparison to a conventional buffer capacitor implementation method and to HVCS. Gate driver architectures and implementations on circuit level are presented in Sect. 4.2. In Sect. 4.3, experimental results verify the gate driver functionality and the calculations of Sect. 4.1. Section 4.4.1 describes limitations of HVES and discusses, which capacitor implementation method in a gate driver is preferred for a particular design target. Finally, the gate driver is compared to the state of the art in Sect. 4.4.3.

---

[1]© 2017 IEEE. Reprinted, with permission, from Seidel, A. and Wicht, B. "Integrated Gate Drivers Based on High-Voltage Energy Storing for GaN Transistors". In: IEEE Journal of Solid-State Circuits, pp. 1–9. ISSN: 0018-9200. DOI: https://doi.org/10.1109/JSSC.2018.2866948.

© The Author(s), under exclusive license to Springer Nature Switzerland AG 2021
A. Seidel, B. Wicht, *Highly Integrated Gate Drivers for Si and GaN Power Transistors*, https://doi.org/10.1007/978-3-030-68940-7_4

4.1   The Concept of HVES

4.1.1   Gate Charge Delivery

The goal addressed in this work is to achieve efficient, reliable, and compact power electronics due to fast switching and highly integrated gate drivers. The integration of all buffer capacitors contributes to this goal by reducing the effect of parasitic inductances, decreasing the number of discrete components and providing higher reliability due to fewer connections (see Sect. 2.1.4). Figure 4.1a shows the equivalent circuit for a driver turn-on event. At turn-on, a voltage drop ΔV_C occurs at C_{Buf}. As, conventionally, ΔV_C needs to be small (typ. < 1 V, Sect. 2.1.2), Eq. (2.3) ($Q_G = C_{Buf} \cdot \Delta V_C$) indicates that only a small amount of charge can be delivered. Chapter 3 introduces the concept of high-voltage charge storing (HVCS) [3], which operates with a much higher ΔV_C to increase Q_G. However, significant energy losses occur in the resistive charging path, as will be further described in Sect. 4.1.3. HVES instead is a resonant concept, which utilizes an inductor L in the gate charge path, as shown conceptually in Fig. 4.1b. The stored energy in C_{Buf}, $E_{CBuf} = 0.5 \cdot C_{Buf} \cdot V_C^2$, is transferred to node V_{GS}, which sees an equivalent energy of $E_{CG} = 0.5 \cdot C_G \cdot V_{GSon}^2$, with V_{GSon} as the nominal on-state gate-source voltage of the power transistor. Due to the resonance, V_C is oscillating and C_{Buf} can discharge completely. This results in higher charge delivery, compared to a conventional buffer capacitor and to the concept of HVCS. The freewheeling diode D_{FW} clamps V_C to ground in order to conduct the current drawn from L while transferring the remaining energy in L to the gate. A rectifier diode D_R prevents a charge backflow out of the gate. Assuming ideal components (diode forward voltages of 0 V, $R_{loop} = 0\,\Omega$), no losses occur. $E_{CBuf} = E_{CG}$ leads to a gate charge delivery of

$$Q_{G,ideal} = \frac{C_{Buf} \cdot V_C^2}{V_{GSon}}. \tag{4.1}$$

$Q_{G,ideal}$ is independent of the inductor L, since for ideal components, L only determines how fast the gate charges, but not the charge quantity. In non-ideal case, a

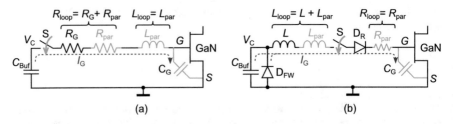

Fig. 4.1 Buffer capacitor implementation methods: (**a**) Conventional and high-voltage charge storing; (**b**) high-voltage energy storing

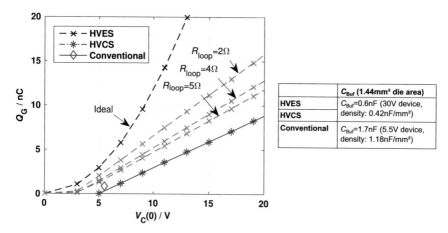

Fig. 4.2 Gate charge delivery versus pre-charge voltage level $V_C(0)$ for various buffer capacitor implementation concepts (HVES, HVCS, and conventional) for a constant layout area of $1.44\,\text{mm}^2$

trade-off has to be found for the sizing of L. A smaller inductance value increases the gate drive speed because the gate current can increase faster. However, at the same time, a small L reduces the resonance effect, as the parasitic resistances in the gate loop get dominant. The case with non-ideal components is discussed further below, and the gate drive speed in particular in Sect. 4.1.2. Figure 4.2 shows the delivered gate charge $Q_{G,\text{ideal}}$ along with the results for various capacitor implementation methods as a function of the initial voltage at V_C. The value of C_G is varied to get a final V_{GS} of 5 V (typical for GaN) at each Q_G point in the diagram. For all methods, the same die area of $1.44\,\text{mm}^2$ is assumed, comprising C_{Buf} and L, which allows a comparison to the experimental results in Sect. 4.3. L is placed on top of C_{Buf} without area penalty. Considering the appropriate voltage ratings, the conventional method results in $C_{\text{Buf}} = 1.7\,\text{nF}$ and for the HVCS and HVES in $C_{\text{Buf}} = 0.6\,\text{nF}$, occupying all the same die area. The conventional buffer capacitor refers to a single point, corresponding to an initial voltage of 5.5 V at V_C and $\Delta V_C = 0.5\,\text{V}$ (see Sect. 2.1.2). Since the energy in C_{Buf} is proportional to V_C^2, a significantly higher gate charge Q_G can be delivered with HVES, compared to the other methods. In contrast to conventional resonant gate drive concepts, V_C can be chosen significantly higher (or lower) than the actual gate drive voltage (see Sect. 2.1.6).

As there is some parasitic resistance R_{par} in the gate charge path, i.e., $R_{\text{loop}} > 0$, resulting in losses and lower energy from C_{Buf}, smaller Q_G is achieved. Q_G can be calculated from the loop equation of the equivalent circuit in Fig. 4.1b:

$$V_L + V_{DR} + V_{R\text{loop}} + V_{GS} - V_C = 0. \tag{4.2}$$

As the current $I_G(t)$ is the same through all components, the loop can be expressed in the Laplace domain with an initial voltage $V_C(0)$ at C_{Buf}:

$$I_G(s) \left(Ls + R_{\text{loop}} + \frac{1}{sC_G} + \frac{1}{sC_{\text{Buf}}} \right) + V_{\text{DR}} = \frac{V_C(0)}{s} \tag{4.3}$$

Rearranging (4.3) and transforming it into the time domain results in

$$I_G(t) = \frac{V_C(0) - V_{\text{DR}}}{L} \cdot \frac{1}{\omega_r} e^{-\frac{R_{\text{loop}}}{2L} t} \cdot \sin(\omega_r \cdot t \cdot h(t)), \tag{4.4}$$

with the natural frequency

$$\omega_r = \sqrt{\omega_o{}^2 - \left(\frac{R_{\text{loop}}}{2L} \right)^2}, \tag{4.5}$$

and the resonance frequency

$$\omega_0 = \sqrt{\frac{1}{LC_{\text{loop}}}}, \tag{4.6}$$

as well as $C_{\text{loop}} = 1/(1/C_G + 1/C_{\text{Buf}})$. (4.4) is the solution for an underdamped system ($\omega_r > 0$). (4.4) holds until V_{GSon} is reached, which corresponds to $I_G(t) = 0$ at $t = \pi/\omega_r$. Hence, the delivered charge Q_G can be calculated by

$$Q_G = \int_0^{\frac{\pi}{\omega_r}} I_G(t)dt \tag{4.7}$$

and

$$V_{\text{GSon}} = \frac{Q_G}{C_G}. \tag{4.8}$$

For small R_{loop}, V_C may swing to negative values. Depending on the gate driver implementation and technology, V_C is clamped one diode forward voltage below ground. For larger R_{loop}, which is often the case for a fully integrated HVES circuit, the clamping diode D_{FW} is hardly conducting. Therefore, its influence is neglected in (4.2) to (4.8).

Figure 4.2 also contains the gate charge Q_G for $R_{\text{loop}} = 2\,\Omega$, $4\,\Omega$ and $5\,\Omega$ with respect to V_C, according to (4.7) and (4.8). L is chosen to be 18.5 nH, corresponding to the experimental results in Sect. 4.3. L can be implemented as coreless inductor on-chip. As L can be placed on top of C_{Buf}, C_{Buf} determines the chip area of L. V_{DR} is set to 0 V, since the diode D_R in this work is implemented as an active rectifier. However, D_R contributes to R_{loop}. Figure 4.2 indicates lower Q_G sensitivity with regard to R_{loop} towards higher values of R_{loop}. The results confirm that the HVES concept is even suitable for an integrated inductor L with a relatively high

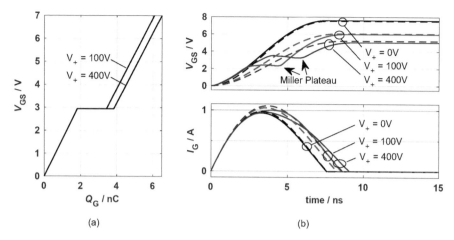

(a)   (b)

Fig. 4.3 (a) Gate charge behavior of a GaN transistor [4] and (b) transistor-level simulations of a GaN transistor from an HVES circuit. V_+ is the drain-source voltage during power transistor off-state

R_{loop}, still providing significantly higher gate charge Q_G compared to the other methods. Especially for a high storage voltage $V_C(0)$, but also for $V_C(0) < 5$ V (low-voltage design), HVES has a superior Q_G-delivery capability. This enables a buffer capacitor integration to drive most of the high-voltage as well as low-voltage/high-current GaN types, currently available in the market.

Equations (4.2) to (4.8) assume a linear gate capacitance C_G (i.e., constant value) with respect to its charging voltage. Figure 4.3a shows a typical non-linear Q_G behavior of a real GaN transistor [4]. The total required gate charge varies due to the Miller effect (see Sect. 2.3). This results in different final gate voltages (Fig. 4.3b), depending on V_+. In order to approximate this settled voltage level by (4.4) to (4.8), equivalent linear capacitances $C_{G,eq}$ can be calculated according to (4.8) with different Q_G values, depending on the Miller charge and the desired V_{GSon}. Figure 4.3b confirms this approximation, since the gate current waveforms I_G is only slightly affected, compared to the I_G waveform for a non-linear capacitance. Nevertheless, the gate driver implementations in Sect. 4.2.1 include a circuit technique comprising a small low-voltage buffer capacitor (C_{DRV}) to compensate gate-voltage variations.

The concept of HVES achieves high gate charge delivery the higher the capacitor voltage $V_C(0)$ is, as shown in Fig. 4.2 for integrated capacitors of the same area. This can be demonstrated by means of a plate capacitor with the well-known equation for its capacitance value

$$C_{plate} = \epsilon_r \cdot \epsilon_o \cdot \frac{A}{d}, \tag{4.9}$$

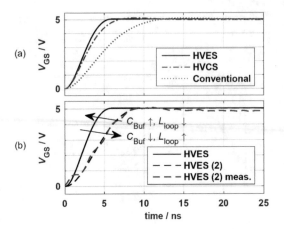

Fig. 4.4 Gate charge speed of the HVES circuit for various L and C_{Buf}

with the area of the electrodes A and the distance d between them. For integrated capacitors, a small chip area A is important. The area of capacitors grows only in one dimension in accordance with its voltage capability. As the gate charge delivery of HVES relates to the energy stored in the buffer capacitor, which is proportional to the square of $V_C(0)$, HVES is in particular beneficial at high voltages.

In power electronics applications, passive components like capacitors or inductors occupy large volume. It could be an idea to apply the concept of HVES somehow, to reduce the volume of some discrete capacitors. HVES requires a high storage voltage, which demands for a high capacitor breakdown voltage. This implies a larger distance d between the capacitor plates. Considering (4.9), A needs to grow proportionally to keep a certain C_{plate} value. Hence, in a first order, the volume of capacitors increases with the power of two referred to its voltage capability. As HVES delivers charge referred to the square of its storage voltage as well, a higher storage voltage is not necessarily more beneficial for discrete capacitors. This is also shown in [5], where an even lower capacitor storage voltage provides a higher energy storage capability.

Nevertheless, the concept of HVES is highly interesting for circuits with discrete buffer capacitors, for which the volume does not count and other advantages of HVES are important. For example, gate drivers with discrete buffer capacitors benefit from its high gate charge speed, which will be introduced in Sect. 4.1.2. Chapter 5 presents high-speed gate drivers based on HVES more suitable with discrete capacitors driving power transistors via large gate loops [6].

4.1.2 Gate Drive Speed

Figure 4.4a shows simulation results demonstrating the gate charging speed of all three capacitor implementation methods: conventional, HVCS, and HVES (see Sect. 4.1.1). All concepts assume the same exemplary gate loop inductance of 6 nH for the typical scenario of Fig. 2.5b. The maximum possible speed relates to the resonance frequency, according to (4.6). This applies to all methods, since all contain a resistance R_{loop}, an inductance L_{loop}, and the capacitors C_{Buf} and C_G in the gate loop (Fig. 4.1). HVES operates with the smallest buffer capacitor C_{Buf} delivering the same gate charge leading to the highest resonance frequency and thus inherently the highest gate drive speed capability. The conventional case and also HVCS require a damping resistor R_G preventing a gate overshoot, which slows down the gate charge process (see Sect. 2.1.4). Assuming ideal circuit components, the concept of HVES exhibits no gate overshoot as the rectifier diode D_R prevents a charge backflow from the gate. The gate overshoot in real circuit is further discussed in Sect. 4.4.1. Figure 4.4b shows the V_{GS} waveform of the HVES. A three times larger inductance of 18.5 nH (waveform "HVES (2)" in Fig. 4.4b) increases the quality factor of the gate loop, resulting in a ~10 % smaller C_{Buf}. The configuration is still ~25 % faster than the conventional gate driver circuit. The V_{GS} measurement shown in Fig. 4.4 (waveform "HVES (2) meas.") confirms this fast gate drive speed. The discrete gate driver in [7] achieves also a very fast gate driving by applying a high voltage at the GaN transistor's gate, decoupled by a capacitor. At turn-on, the driven GaN gate injection transistor (GIT) device (see Sect. 2.2.2) clamps the excessive charge preventing a gate overshoot. A GaN transistor with isolated gate (see Sect. 2.2.2) would require an additional clamping structure that needs to be accurate enough. Furthermore, clamping structures dissipate the energy stored in the gate loop inductance in comparison to resonant gate driver concepts leading to a lower gate driver efficiency (see Sect. 2.1.7).

A smaller C_{Buf} is an option to increase the gate drive speed. As this would reduce the gate charge delivery, a higher initial voltage $V_C(0)$ can be chosen for compensation. Another possibility to compensate the gate charge reduction is a lower R_{loop} (see Fig. 4.2), for example, by providing a switch with a lower $R_{DS,on}$ in the gate loop.

Figure 4.5 shows LTspice simulations of the HVES, varying $V_C(0)$ from 14.4 V to 100 V, (a) with and (b) without the free-wheeling diode D_{FW}. C_{Buf} is adapted to charge the gate ($C_G = 1.6$ nF) to 5 V. As expected, Fig. 4.5a shows an increasing gate drive speed with a higher $V_C(0)$. In theory, an unlimited gate charge speed can be achieved by increasing the gate loop resonance frequency (see (4.6)). The high gate charge speed of the HVES concept results from the small total gate loop capacitance C_{loop} consisting of the series connection of C_G and C_{Buf}. However, the series connection of C_G and C_{Buf} takes no effect anymore when the free-wheeling diode D_{FW} starts to conduct and shortens C_{Buf}. Hence, the concept of HVES can only speed up the V_{GS} waveform compared to the conventional case before the free-wheeling conducts the gate current, as the gate-current and V_C waveforms indicate.

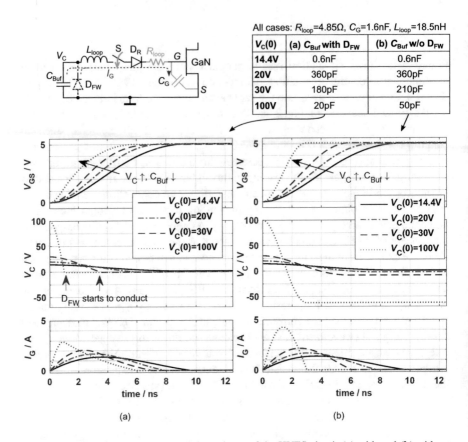

Fig. 4.5 $V_C(0)$ and gate charge speed dependency of the HVES circuit (**a**) with and (**b**) without free-wheeling diode D_{FW}

D_{FW} gets only active when the quality factor of the gate loop is high enough such that V_C discharges below zero. In other words, this kind of gate drive speed limitation applies, in particular, if C_{Buf} and R_{loop} are small and L is large (see Sect. 2.1.4). For comparison, Fig. 4.5b shows simulations at the circuit shown in Fig. 4.5 on the left, but without D_{FW}, to avoid this gate drive speed limitation. This allows node V_C to swing negative and C_{Buf} takes effect in the gate loop during the whole gate charge process. This speed-optimized gate drive circuit has the disadvantage that C_{Buf} needs to be designed larger, as further described below in this section. At a $V_C(0) = 100$ V, a V_{GS} rise time of 1.7 ns is achieved, which is 50 % faster than the circuit with D_{FW} (5.8 ns rise time) even with a higher value of C_{Buf}. Compared to a conventional gate driver with a rise time of 13 ns, this is seven times faster (see Fig. 4.4). C_{Buf} needs to be designed larger without D_{FW} because at the end of a transistor turn-on event some energy remains in the negatively charged C_{Buf}. This contradicts to a fully integrated gate driver. Furthermore, it is more challenging to design integrated circuits that cope with negative voltages.

Depending on the design goal, the gate drive circuit can be optimized for a smaller size of C_{Buf} or faster gate charging. Section 6.3 shows a gate driver implementation based on HVES optimized for gate drive speed with partially discrete components. This gate driver architecture allows the buffer capacitor to take effect during the whole gate charge phase without negative voltages on chip.

4.1.3 Efficiency Considerations of Buffer Implementation Methods

This section describes the efficiencies of the different capacitor implementation concepts. As described in Sect. 2.1.7, the efficiency of a conventional gate driver turn-on event is ideally 50 %. Only resonant gate driver topologies can achieve higher values, in theory 100 % (see Sect. 2.1.7).

The efficiency of a driver turn-on event of the HVCS and HVES concepts can be calculated as the ratio of the energy in the charged gate capacitance C_G (E_{out}) to the incoming energy from C_{Buf} (E_{in}):

$$\eta = \frac{E_{out}}{E_{in}} = \frac{0.5 \cdot C_G \cdot V_{GSon}^2}{0.5 \cdot C_{Buf} \cdot (V_C(0)^2 - V_{C,min}^2)} = \frac{C_G \cdot V_{GSon}^2}{C_{Buf} \cdot (V_C(0)^2 - V_{C,min}^2)}.$$
(4.10)

$V_{C,min}$ is the remaining voltage at C_{Buf} after the turn-on process. In case of a low quality factor of the gate loop, C_{Buf} is not fully discharged and $V_{C,min}$ can be $> 0\,V$ (see also Sect. 4.1.2). For the efficiency of the HVCS circuit, $V_{C,min}$ is always the final on-state gate-source voltage V_{GSon}.

In theory, the HVES concept achieves 100 % efficiency, due to its resonant topology. If (4.10) is considered, this occurs if $E_{out}/E_{in} = 1$. $V_{C,min}$ is $0\,V$ in this case. Hence, the ratio C_{Buf}/C_G equals the ratio $V_{GSon}^2/V_C(0)^2$. However, the parasitic resistance R_{par} in the gate charge path reduces the overall efficiency in (4.10). For the measurements in Fig. 4.6, R_{par} of the HVES-based gate driver comprises L (1.5 Ω), S (1.5 Ω), the active rectifier (0.35 Ω), and the internal gate resistance of the GaN transistor (1.5 Ω). In Fig. 4.6, efficiency measurements at the gate driver based on HVES are compared to calculated values. The efficiency of the HVES concept is 1.3 times superior to HVCS.

For theoretical efficiency calculations of the HVES circuit, the value for $V_{C,min}$ is typically unknown. In this case, the efficiency of the HVES circuit can be calculated based on the fact that the energy taken from the input is equal to the sum of the energy E_{out}, delivered to the gate, and the internal loss $E_{loss,HVES}$.

$$\eta_{HVES} = \frac{E_{out}}{E_{in}} = \frac{E_{out}}{E_{out} + E_{loss,HVES}} = \frac{1}{1 + \dfrac{E_{loss,HVES}}{0.5 \cdot C_G \cdot V_{GSon}}}.$$
(4.11)

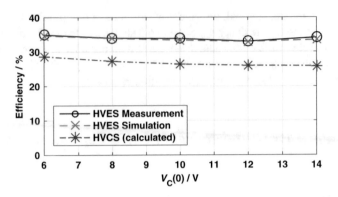

Fig. 4.6 Efficiency comparison of the buffer capacitor implementation concepts HVES and HVCS

The energy loss

$$E_{\text{loss,HVES}} = E_{\text{out}} - E_{\text{in}} = \int_0^{\frac{\pi}{\omega_r}} I_G^2(t) \cdot R_{\text{par}} dt = R_{\text{par}} \cdot \int_0^{\frac{\pi}{\omega_r}} I_G^2(t) dt \qquad (4.12)$$

is derived from the gate current $I_G(t)$ through R_{par}, based on (4.4) and (4.5).

The loss calculations in (4.10) and (4.12) consider only the charge delivery phase, i.e., the discharging of C_{Buf}. However, according to Sect. 2.1.7, in case of a non-resonant recharging of C_{Buf} from a voltage source, the recharge efficiency is only 50 % (assuming charging from zero to $V_C(0)$). To get the overall gate driver turn-on efficiency, the recharge efficiency needs to be multiplied with η (of HVES). In order to minimize the recharge losses, C_{Buf} can be charged resonantly or via a circuit with current source behavior at its output (Sect. 2.1.7). For example, an isolated gate driver supply comprising a transformer can be designed such that it has a current source behavior [8]. Section 6.4 proposes a high-side gate driver supply circuit, which may reduce the C_{Buf} recharge losses as well.

4.1.4   Bootstrap Capacitor Voltage Clamping

Especially in application with GaN transistors, precautions are required to protect the bootstrap capacitor from overcharging, delivering a precisely regulated gate drive voltage (see Sect. 2.4.2). The measurement results in Fig. 4.7 show that the gate voltage of the proposed HVES concept is not very sensitive on V_{HV} variations [1]. Rearranging (4.1) to

$$V_{\text{GSon}} = V_C(0) \cdot \sqrt{\frac{C_{\text{Buf}}}{C_G}} \qquad (4.13)$$

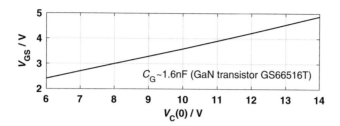

Fig. 4.7 Measurement of V_{GS} versus the initial voltage $V_C(0)$ at the HVES circuit

shows that V_{GSon} is proportional to $V_C(0)$. This equation is only valid for an ideal HVES storing circuit, without losses. Nevertheless, the measurements confirm the equation in terms of the proportional relationship between V_{GSon} and $V_C(0)$.

This relaxes the demand for protection against violation of the maximum gate-voltage ratings of the GaN transistor.

4.1.5 Design Scenarios with HVES

HVES enables three C_{Buf} design scenarios.

1. **Minimum C_{Buf} area.** A smaller C_{Buf} can be integrated more easily to reduce the number of external components. No external components would be necessary anymore supporting co-integration of gate driver and GaN transistor on the same die (see Sect. 2.1.4). Fully integrated GaN transistor half- or full-bridges would save several external buffer capacitors. This is especially advantageous for bipolar or multi-level gate drivers, which require two or more buffer capacitors for every implemented gate driver. Depending on technology, the high-voltage capacitor C_{Buf} as well as the inductor can be implemented on-chip without layout area penalty, on top of the low-voltage capacitor C_{DRV} (see Figs. 4.9 and 4.17).
2. **C_{Buf} placed off-chip.** This allows to adjust the gate charge speed by adjusting C_{Buf} and $V_C(0)$ (see Sect. 4.1.2), e.g., for electromagnetic compatibility reasons. Smaller C_{Buf} leads to higher speed and vice versa. Consequently, $V_C(0)$ needs to be adapted to deliver the same Q_G (Fig. 4.2). This can be an effective alternative to the conventional method, which uses a gate resistor R_G to adapt the speed [9]. In particular, this can be beneficial for an integrated gate driver with GaN transistor on the same die, where R_G is not accessible anymore.
3. **Large L_{loop}.** The gate charge speed characteristics of HVES (see Sect. 4.1.2) allow for fast gate driving, even with large parasitic gate loop inductances. Hence, the gate driver can be placed in large distance to the GaN transistor. Chapter 5 presents gate driver architectures and implementations based on HVES, which are capable of driving large gate loops.

Fig. 4.8 Concept of the gate driver based on HVES with bipolar gate voltage operation

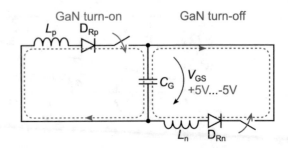

4.1.6   Bipolar Gate Drive Operation

According to Sect. 4.1, the concept of HVES is based on a resonance circuit that oscillates half a period to resonantly charge the gate capacitance C_G. For a unipolar gate drive operation, C_G is charged up to a positive voltage to turn on the power transistor. Its energy comes from C_{Buf} (Fig. 4.1). For turn-off, the energy in C_G is dissipated via the gate driver pull-down transistor. Figure 4.8 shows the concept of a bipolar gate drive operation, assuming ideal components. Before turn-on, C_G is charged negatively and vice versa for turn-off. Hence, the bipolar gate drive operation utilizes the initially stored energy in C_G for each driver turn-on/-off event. The energy in C_G oscillates from the positive to the negative polarization to turn off and on the transistor, respectively. The bipolar gate drive operation requires double the amount of charge in every gate driver switching event (charging C_G from negative to positive V_{GS} and vice versa). The gate driver would have ideally 100 % efficiency. For a real driver implementation, the dissipated energy, caused by non-ideal components, needs to be added to the resonance circuit. This benefit has more effect on an HVES resonance circuit with a high quality factor (see Sect. 2.1.4). More charge flows through the parasitic gate loop resistance, leading to more losses, compared to the unipolar gate drive operation. With a lower quality factor, the bipolar gate drive operation may require an even larger C_{Buf}.

de Vries [10] presents a similar resonant gate driver architecture with oscillating energy in the gate loop, leading to a bipolar gate drive scheme. This architecture requires a constant voltage at C_{Buf}. Hence, the capacitor is not integrated and does not apply HVES with benefits like higher gate drive speed (see Sect. 4.1.2).

To consider a bipolar gate drive operation in the equations (4.2) to (4.8), the absolute value of the initial gate-source voltage $V_{GS}(0)$ needs to be added to the initial buffer capacitor voltage $V_C(0)$, i.e., $V_C(0) := V_C(0) + V_{GS}(0)$.

4.2 Gate Driver Implementation

4.2.1 Gate Driver Architectures

Driver Architecture for Unipolar Gate Voltage Figure 4.9 shows the implementation of the gate driver based on HVES for a unipolar gate voltage [1]. It comprises a fully integrated HVES circuit consisting of an LC-tank (L_{HVp}, C_{HVp}) with C_{HVp} charged to a high voltage V_{HVp} (typically around 15 V) over the "V_{HV} Control" block [1] from V_{SUP}. If the gate control signal DRV_{INp} is activated (timing diagram in Fig. 4.9), M_{HVp} turns on, via the "M_{HV} Control" block [1], and C_{HVp} discharges over L_{HVp} and the active rectifier [1]. This delivers a current pulse $I_{G,source}$ with the required charge Q_G onto the gate of the GaN transistor. During GaN turn-on, V_{HVp} is disconnected from the driver supply V_{SUP} by the "V_{HV} Control" block to prevent an additional energy flow into C_{HVp} and, consequently, GaN gate-source overvoltage due to excessive charging of its gate capacitance C_G. Triggered by the gate signal at node G, a low voltage buffer in parallel, comprising MN_p, MN_p, keeps the GaN switch safely connected to the low-voltage capacitor C_{DRV} at the end of the gate charge transition. Furthermore, C_{DRV} compensates variations of the gate charge Q_G. For example, an excessive charge $Q_{ov} = 0.5\,nC$ for an operation range from $V_+ = 100\,V$ to $400\,V$ (Fig. 4.3) would lead to a gate overshoot $\Delta V_{GS} \sim 1\,V$ without C_{DRV}. In contrast, $C_{DRV} = 3.6\,nF$ and $C_G = 0.62\,nF$ result in a $\Delta V_{GS} = Q_{ov}/(C_{DRV} + C_G) = 120\,mV$. Even at $V_+ = 0\,V$ (lowest required Q_G), V_{GS} varies by less than 0.5 V. After a delay, M_{HVp} turns off and the "V_{HV} Control" block turns on. The blocks "M_{HV} Control" and "V_{HV} Control" are discussed in more detail in Sect. 4.2.5. Recharging C_{HVp} immediately after gate driver turn-on enables a recharge over most of the switching period, even while the driver is in the on-state. This particular advantage of the proposed gate driver supports short off-times of the

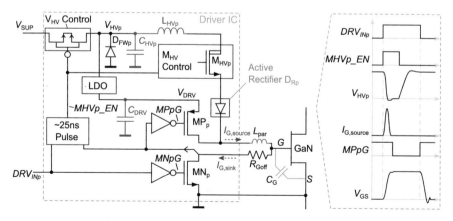

Fig. 4.9 Proposed gate driver based on HVES, providing unipolar gate drive voltage

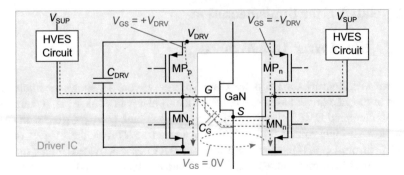

Fig. 4.10 Proposed gate driver based on HVES, providing bipolar and three-level gate drive voltages

GaN transistor. The GaN transistor is turned off via MN_p with $I_{G,sink}$, which can be adjusted by an optional external gate resistor $R_{G,off}$.

A design rule for the minimum value of L_{HV} results from the voltage divider formed by L_{HVp} and L_{par} at the beginning of the GaN turn-on event (see Fig. 4.9). If L_{HVp} is smaller than $L_{par}(V_{HVp}(0)/V_{DRV} - 1)$, an undesired current flows into V_{DRV} via MN_p. For example, $L_{par} = 6\,nH$, $V_{HVp} = 15\,V$, and $V_{DRV} = 5\,V$ yield $L_{HVp} \geq 12\,nH$. The smaller the L_{par}, the broader the range of recommended values for L_{HVp}.

Driver Architecture for Bipolar and Three-Level Gate Voltage To obtain a gate driver with a bipolar and a three-level gate drive voltage, the unipolar gate driver in Fig. 4.9 is applied two times, at the gate G and at the source S, leading to a full-bridge architecture comprising MN_p, MN_p, MN_n, and MN_n as shown in Fig. 4.10. Depending on the bridge polarity, a positive, negative, and zero gate-source voltage V_{GS} is applied to the GaN device. The different current paths in Fig. 4.10 show the fast gate charge/discharge paths from the HVES circuits, and the paths through the full bridge transistors keeping the device constantly at V_{DRV} level. Perez et al. [11] presents a gate driver with a full-bridge architecture, with the aim of optimizing the switching behavior of IGBT and to decrease the number of buffer capacitors to one. Since it is a discrete buffer capacitor, this architecture has four bond wires in the gate loop. With HVES instead, only two bond wires are required. Section 4.2.2 presents the details on a gate driver implementation providing a three-level gate voltage that can be supplied with a non-isolated or isolated gate driver supply.

Driver Architecture for Bipolar Gate Voltage with Only One HVES Circuit In order to save significant die area, the "HVES circuit" blocks in Fig. 4.10 can be implemented only once, such that gate and source share the same HVES. Only the "M_{HV} Control" block and the active rectifier (Fig. 4.9) are needed twice. This architecture option limits the minimal driver on-/off-state duration as the recharge process of the high-voltage capacitor C_{HVp} requires some time. This duration can be tens of ns depending strongly on the technology and design target. Furthermore,

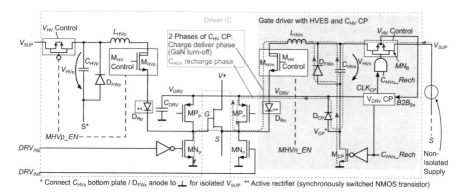

Fig. 4.11 Gate driver circuit for an isolated and non-isolated supply, supporting bipolar and three-level gate drive operation

if HVES is shared, the driver cannot be supplied by a bootstrap circuit, since the C_{HVp} capacitor needs to be recharged after a turn-on event.

4.2.2 Gate Driver Implementation for a Non-isolated and an Isolated Supply

As explained in Sect. 2.4.1, a non-isolated driver supply recharges only buffer capacitors that are referred to the GaN source. Applied to the bipolar and three-level gate driver with a full-bridge architecture (Fig. 4.10), all buffer capacitors can be recharged only during GaN on-state, as the chip ground is connected to the node S via transistor MN_n. However, in some applications, the driver operates with too short driver on-state phases to recharge the buffer capacitors sufficiently. Similarly, if the gate driver is supplied by a bootstrap circuit, the supply is only available during GaN off-state (Fig. 2.14).

Based on Figs. 4.9 and 4.10, Fig. 4.11 shows a modified driver implementation with its timings in Fig. 4.12 that allows to recharge all buffer capacitors even during driver off-state. To turn on the GaN switch, the driver control signal DRV_{INp} initiates a turn-on event, similar to the driver in Fig. 4.9. In case of a bipolar driver operation, DRV_{INn} turns "high" at the same time, switching off MN_n and turning on MN_n to discharge the source S. For driver turn-off, G discharges via MN_p, controlled by the driver signal DRV_{INp}. Triggered by DRV_{INn}, the high-voltage transistor M_{HVn} initiates a resonant current pulse from C_{HVn} over L_{HVn}, charging node S. The current path is closed via MN_p and the low-voltage charge pump transistor M_{CP}, which is in on-state during driver turn-off. Finally, MN_n connects S to V_{DRV}. In three-level mode, DRV_{INn} turns "high" before the next driver turn-on event, discharging node S via MN_n.

Fig. 4.12 Timings of the
gate driver circuit in Fig. 4.11

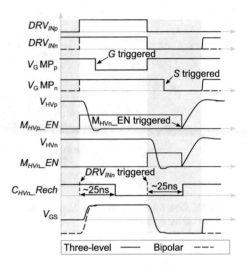

In case of an isolated supply (supply ground referred to chip ground), C_{HVp}/C_{HVn} can be connected to the chip ground. In order to recharge C_{HVp}/C_{HVn} from V_{SUP} (typ. ~15 V), both "V_{HV} Control" blocks turn on and the current path closes over the capacitors and the chip ground. In case of a non-isolated supply (supply ground referred to the source S), the source node S is connected to V_{DRV} via MN_n, during driver off-state (negative gate level). As the capacitor's C_{HVp}/C_{HVn} bottom plates are connected to the chip ground, the charging current would flow through C_{DRV} and MN_n to S discharging C_{DRV}. Therefore, the bottom plate of C_{HVp} can be connected to the source node S of the GaN transistor. However, for C_{HVn} this is not possible, because it needs to have a chip ground connection during driver turn-off.

For this reason, the implemented CHV charge pump ("C_{HV} CP"), comprising M_{CP} and D_{CP}, reuses C_{HVn} as a charge pump capacitor. The "C_{HV} CP" operates in two phases (see charge flow in Fig. 4.11). In the "charge deliver phase, " at driver turn-off, node G is shorted to chip ground while M_{CP} is in the on-state in order to charge node S. The resonant characteristics of the gate driver with HVES causes the voltage at node S to exceed V_{DRV} such that C_{DRV} recharges via MN_n. The same C_{DRV} recharge scheme applies on the gate side at driver turn-on. The second charge pump phase ("C_{HVn} recharge phase") occurs after driver turn-off. M_{CP} turns off and C_{HVn} is recharged via D_{CP} and MN_n directly to the source.

The gate driver and the 'C_{HV} CP" (Fig. 4.11) deliver the charge once per driver switching cycle. In some power applications, it is necessary to provide a negative gate voltage during long driver off-state phases or at start-up of the particular application. To supply V_{DRV}, an additional charge pump (V_{DRV} CP) can be activated (Fig. 4.13) that needs to deliver only the static current drawn from V_{DRV}. This charge pump works in two phases. During the first phase, C_{CP} is recharged from V_{SUP} (typ. 15V) over the first transistor of the back-to-back switch (Fig. 4.11) on the GaN transistor's source side. The back-to-back switch prevents a discharge of the C_{HVn}

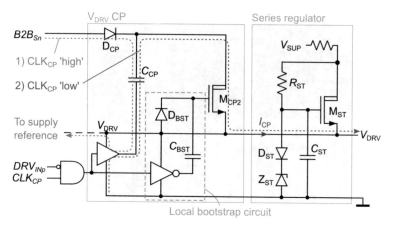

Fig. 4.13 Circuit implementation of the "V_{DRV} CP" with start-up

capacitor during charge pump phase 2. The diode D_{CP} prevents a discharge of V_{DRV} over the body diode of M_{CP2} if the gate driver supply voltage is not connected, e.g., during turn-on with bootstrap supply. During the second "V_{DRV} CP" phase, C_{CP} discharges to V_{DRV}. The high voltage drop at C_{CP} delivers a large amount of charge due to the effect of HVCS [3]. The provided current can be approximated by

$$I_{CP} = (V_{SUP} - V_{DCP} - V_{DRV}) \cdot C_{CP} \cdot f_{CP} \qquad (4.14)$$

with the forward voltage V_{DCP} of the diode D_{CP} and the switching frequency f_{CP} of the signal CLK_{CP}. For example, $C_{CP} = 50\,pF$ provides an average I_{CP} of 0.5 mA at $f_{CP} = 1\,MHz$, $V_{SUP} = 15\,V$ and $V_{DRV} = 5\,V$. The "V_{DRV} CP" efficiency equals to that of a comparable linear regulator (LDO), since its average input current is equal to the average output current I_{CP}. In contrast to an LDO, this type of charge pump increases not the efficiency, but it is able to pump charge from one to another voltage reference in a very area-efficient manner.

4.2.3 Operation as High-Side Gate Driver

This section describes the gate drivers, shown in Figs. 4.9 and 4.11, operating as high-side gate drivers. High-side gate drivers experience coupling currents between the gate driver and its control unit on the low-side, in contrast to low-side gate drivers (see Sect. 2.4.5). Especially, high-side gate drivers with small buffer capacitors, which is typically the case for drivers based on HVES, require clamping structures to protect the capacitors against discharge or overcharge.

Figure 4.14 shows a simplified circuit of the three-level gate driver (Fig. 4.11) in a half-bridge topology, (a) with an isolated and (b) with a non-isolated driver supply.

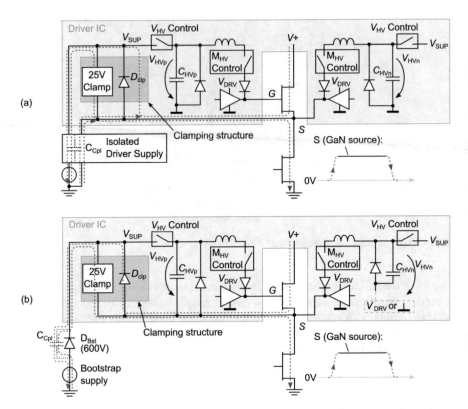

Fig. 4.14 Coupling currents in a high-side gate driver operation supplied by (**a**) isolated gate driver supply and (**b**) bootstrap supply

For the unipolar gate driver topology in Fig. 4.9, the depicted coupling currents are identical. During rising voltage slopes of the source node S of the high-side GaN transistor, V_+ charges the coupling capacitance C_{cpl} pulling down V_{SUP}. The "V_{HV} Control" block prevents a discharge of the capacitors C_{HVp} or C_{HVn}. The integrated clamping diode D_{clp} prevents V_{SUP} from falling below the potential of node S, that is at chip-ground level at turn-on. Without D_{clp}, substrate diodes in the "V_{HV} Control" block would become conductive with the risk of failures and latch-up. During falling voltage slopes of the source node S, the coupling capacitance C_{cpl} is discharged while charging node V_{SUP}. As V_{SUP} is clamped at 25 V with respect to the source node, no voltage-rating violations occur at the "V_{HV} Control" block. The clamping circuit consists of a high-voltage NMOS transistor. A resistor pulls down its gate. In case of an overvoltage, a stack of zener-diodes pulls up the gate (like the active 25 V-clamp in Fig. 4.14). Nevertheless, a gate driver supply with a low C_{cpl} is crucial for minimizing coupling currents.

4.2.4 GaN GIT Support

The proposed three-level driver designs in Figs. 4.9, 4.10, and 4.11 also support gate injection transistor (GIT) devices, which have special demands regarding the gate drive circuit (see Sect. 2.2.3). The HVES circuit provides large gate currents for fast GaN turn-on/-off events. During GIT on- and off-state, MN_p/MN_n and MN_n/MN_p, respectively, provide a low-resistance connection to C_{DRV} buffering the gate voltage during interferences like Miller coupling. The linear regulator (LDO) in Fig. 4.9 and the series regulator in Fig. 4.13 deliver a DC gate current I_G in the mA-range during driver on-state limiting the dc current that can be drawn from V_{DRV}. For non-GIT GaN devices, the LDO acts as a voltage regulator for V_{DRV}.

4.2.5 Driver Sub-circuits

4.2.5.1 V_{HV} Control

The "V_{HV} Control" circuit in Fig. 4.15 controls the recharging of the buffer capacitors C_{HVx} in Figs. 4.9 and 4.11. It connects or disconnects C_{HVx} from the driver supply V_{SUP} via the 30V DMOS transistors MN_{B1} and MN_{B2}. The back-to-back configuration of MN_{B1}, MN_{B2} prevents a discharge of C_{HVx} into V_{SUP} via the body diode of MN_{B1} in case V_{SUP} drops below V_{HVx}. At low switching frequencies and at start-up, C_{HVx} is charged directly via a bypass resistor $R_1 = 10\,k\Omega$. C_{HVx} is disconnected from V_{SUP} at turn-on/-off of the GaN switch. The recharge mode ($MHV\_EN = 0$) is entered. In this case, MN_3 is turned-off and MP_1 connects V_{G12} to V_{SUP} via a bootstrap diode D_3, controlled over C_1. D_3 prevents a charge backflow to V_{SUP}, if V_{G12} exceeds V_{SUP}.

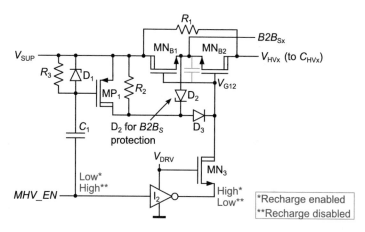

Fig. 4.15 "VHV Control" block

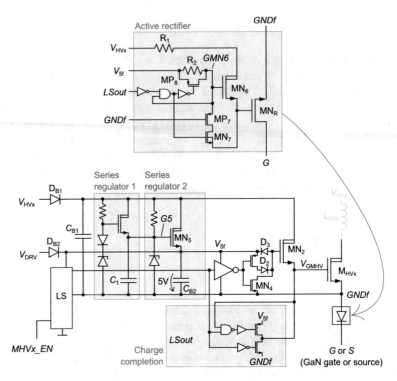

Fig. 4.16 MHVx control and supply

4.2.5.2 "M_HVx Control" and Active Rectifier

The control and supply circuit for the transistors M_{HVx} (Fig. 4.16 bottom) triggers the HVES current pulse in Figs. 4.9 and 4.11. M_{HVx} turns on via MN_2, controlled by the level shifter LS. D_2 prevents discharging of the MN_2 gate capacitance when node V_{GMHV} rises. D_3 ensures a 5 V-overvoltage protection for MN_4. As MN_2 turns off, the "Charge Completion" block pulls V_{GMHV} to its final level V_{5f} (5 V). The floating supply V_{5f} is generated by two cascaded series regulators. The small "Series Regulator 1" rapidly resets capacitor C_1 of the larger "Series Regulator 2" to its target value at C_{B1} recharge. C_{B1} and C_{B2} are stacked devices. The high voltage drop at C_{B1} results in a high charge allocation. Thus, this circuit is based on the concept of HVCS (Chap. 3), occupying a small die area. The low voltage bootstrap capacitor C_{B1} cannot be overcharged, since the "Series Regulator 2" stops charging C_{B1} at its defined voltage level. Hence, this circuit is very insensitive to bootstrap supply voltage variations. In the output stage in Fig. 3.3, this can be addressed by not implementing MP_1, as described in Sect. 3.2. In contrast to the gate driver output stage in Fig. 3.3, the V_{5f} voltage rail is mainly supplied from V_{HVx} via the series

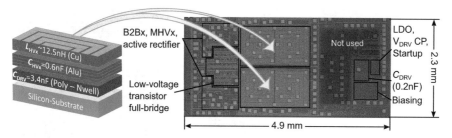

Fig. 4.17 Microphotograph of the implemented gate driver IC

regulators and not from V_{DRV} via the bootstrap capacitor D_{B2}. Therefore, lower peak currents are drawn from V_{DRV}. As a disadvantage, the bias currents through the series regulators in Fig. 4.16 might have a considerable efficiency impact that is less dominant in fast switching applications.

The low-voltage transistor MN_R acts as an active rectifier (Fig. 4.16 top) that turns off via MN_6 and R_1, once V_{HVx} falls below the GaN gate voltage G at each turn-on transition. R_1 causes a short delay to keep MN_R ideally active until its drain current reaches zero. The gate of the cascode MN_6 is protected by MP_7. When V_{HVx} rises, the cascode MN_7 turns off and $GMN6$ is pulled to V_{5f} via R_2. At the end of the switching cycle, M_{HVx} gets turned off ($LSout$ = "low"). MP_8 shorts R_2 for a faster activation of MN_6 and MN_7 pulls the gate of MN_R fully to V_{5f}.

4.2.5.3   Level Shifter

The level shifter (LS) for the "M_{HVx} Control" circuit in Fig. 4.16 is a low-voltage version of the level shifter presented in Sect. 3.3.3 as the floating ground of the level shifter (source of M_{HVx}) experiences maximum voltages of only $\sim$5 V. Hence, the transistors MN_1, MN_2, MP_1, MP_2 in the level shifter circuit (Fig. 3.6) can be implemented as 5 V-rated transistors.

4.3   Experimental Results

The gate driver circuit of Fig. 4.11 has been manufactured in a 0.18 μm BCD technology (Fig. 4.17). $2\times$ L_{HVx} = 12.5 nH in 3 μm thick copper are stacked onto $2\times$ C_{HVx} = 0.6 nF without layout area penalty on top of C_{DRV} = 3.4 nF, covering an area of 1.44 mm^2. The total active die area is approximately 5.2 mm^2.

Figure 4.18 shows an example of the experimental assembly of the three-level gate driver on a ceramic substrate, driving a bare die GaN transistor as part of a half bridge module, without the need for any external components.

Fig. 4.18 Three-level gate driver on ceramic substrate as part of a GaN half-bridge module

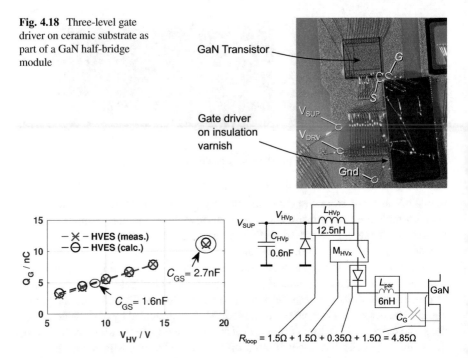

Fig. 4.19 Measured gate charge delivery of HVES compared to calculated values

To confirm the calculations of Sect. 4.1.1, the Q_G delivery of the proposed gate driver with unipolar gate drive operation (using a single chip side) was measured at a 1.6 nF GaN gate and also for a capacitor simulating a larger gate of $C_G = 2.7$ nF. The measurements in Fig. 4.19 show gate charges up to 11 nC and match very well with the calculated values based on (4.7).

Figure 4.20 shows a measurement to verify the basic gate driver function and its driving capability in case of a bipolar gate drive scheme. A capacitor $C_{GS,add}$ was added to the gate-source terminals of the GaN transistor GS66508T (GaN Systems) that artificially increases the gate capacitance. A maximum gate charge of $Q_{max} = 11.6$ nC results from the gate voltage swing ΔV_{GS} charging the total gate capacitance $C_G + C_{GS,add}$. The peak gate drive currents of $I_{G,source,max} = 1.5$ A and $I_{G,sink,max} = 1.3$ A were extracted from the V_{GS} waveform with $I_{Gsource/sink,max} = SR \cdot (C_G + C_{GS,add})$ with the slew rate SR of the rising and falling edge, respectively.

The double pulse test in Fig. 4.21 shows a GaN GIT in three-level mode at a switching voltage $V_+ = 100$ V and $I_L = 10$ A load. The driver IC is operated from an isolated supply. The enlarged V_{GS} waveforms show the Miller plateau of the turn-on and turn-off events that occur during fast V_{SW} transients of > 80 V/3 ns and > 80 V/2.8 ns for the rising and falling slope, respectively.

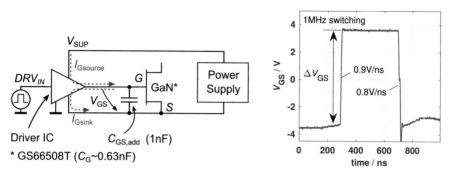

Fig. 4.20 Maximum gate charge and current characterization of the gate driver

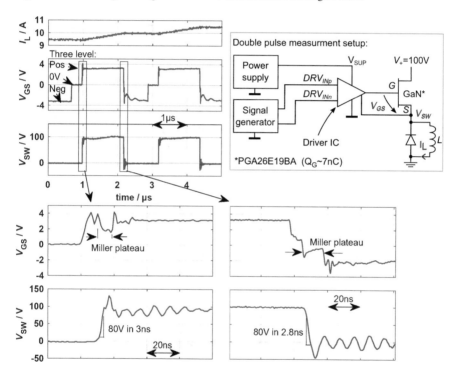

Fig. 4.21 Transient measurements of the gate driver with three-level gate voltage operation

Figure 4.22a shows the benefit of the activated "V_{DRV} CP" with a charge pump switching frequency of 1.2 MHz supporting long driver off-state phases. As the driver has some additional current consuming blocks at V_{DRV} for research purposes, the pumping capacitor $C_{CP} = 270\,pF$ (see Fig. 4.13) is implemented as an external device to deliver sufficient charge. According to (4.14), the C_{CP} provides

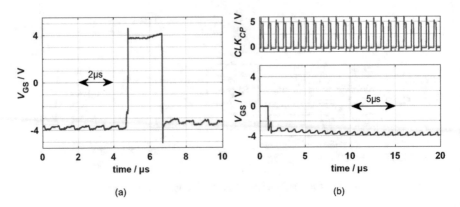

Fig. 4.22 V_{GS} measurements with (**a**) "V_{DRV} CP" enables long driver off-state phases, and (**b**) 'V_{DRV} CP' provides a negative V_{GS} at gate driver start-up

a current I_{CP} of ~2.4 mA from a $V_{SUP} = 12$ V. Figure 4.22b shows a driver start-up measurement. Once the series regulator (Fig. 4.13) provides V_{DRV}, the signal DRV_{INn} (Fig. 4.11) resets to "high" turning V_{GS} to a negative gate voltage. The "V_{DRV} CP" takes over to supply V_{DRV}. In case of a three-level gate drive scheme, no negative gate-source voltage is applied to the GaN transistor before turn-on. Hence, this start-up procedure and also the "V_{DRV} CP" are not necessary.

4.4   State-of-the-Art Comparison and Limitations

This section summarizes the applications for which the individual capacitor implementation methods offer advantages and describe limitations of HVES. Furthermore, the gate drivers based on HVES are compared to the state of the art.

4.4.1   Limitations of Gate Drivers Based on HVES

Higher Gate Drive Speed by Increasing Initial V_C In the implementations in Figs. 4.9 and 4.11, $V_C(0)$ and C_{Buf} are referred to V_{HVx} and C_{HVx}, respectively. V_{HVx} is limited by maximum voltage ratings of the technology. Components in the HVES circuit get larger due to higher voltage ratings, like the transistors M_{HVx}, the back-to-back switches in the "V_{HV} Control" block (Fig. 4.15) and the high-voltage capacitor C_{HVx}. Furthermore, higher gate current requires larger devices in the gate charge path. Therefore, the available die area limits the gate drive speed.

Table 4.1 Performance comparison of the three buffer capacitor implementation methods

| | Buffer capacitor implementation | | |
	Conventional	HVCS	HVES
Size C_{Buf}	–	o	+
Gate drive speed	–	o	+
Small L_{loop}	o	+	–
Efficiency	+	–	+
Complexity	+	o	–

Gate Overshoot The output capacitance of the gate driver leads to more gate overshoot/undershoot or even oscillations at the GaN transistor as some charge flows back into the driver. This capacitance is mainly determined by the rectifier diode D_R (D_{Rp}/D_{Rn} in Figs. 4.9 and 4.11). However, the gate overshoot/undershoot is very low for a small gate loop inductance between gate driver and power transistor. Chapter 5 describes gate driver based on HVES designed for large gate loops. Possible precautions against this issue are discussed in Sect. 5.3.

Minimum Size of L_{HVx} A higher gate drive speed can be achieved by a smaller integrated inductor L_{HVx} in the HVES circuit (see Sect. 4.1.2). As described in Sect. 4.2.1, the minimum size of L_{HVx} is limited to prevent voltage rating violations of the gate driver output. This is not a limit of the HVES concept itself but depends on the gate driver architecture. The gate driver based on HVES designed for large gate loops, described in Chap. 5, does not have this limitation. Furthermore, a small L_{HVx} leads to a low quality factor of the gate loop. If the quality factor is too low, the concept of HVCS might be more beneficial, which is discussed further in Sect. 4.4.2.

4.4.2 HVES Versus HVCS and Conventional Gate Drivers

Table 4.1 shows a comparison of the three buffer capacitor implementation concepts (see also Sect. 4.1). This assessment may vary depending on the individual application and technology. For very small gate charges Q_G, the conventional approach requires the smallest die area because it has the lowest circuit complexity and comprises no high-voltage components. For a higher Q_G, the concept of HVCS can significantly reduce the die area, as shown in Chap. 3 [3]. For much higher charge Q_G values, HVES requires the lowest buffer capacitor size, and, even with the highest complexity, it occupies the overall smallest die area. Furthermore, it has the best achievable efficiency (see Sect. 4.1.3), which is in particular important for high Q_G delivery when the absolute amount of power dissipation is not negligible anymore. Another advantage of the HVES concept is the superior gate drive speed (see Sect. 4.1.2). HVCS has almost the same gate drive speed capability. Section 4.1.5 describes three design scenarios that are beneficial for gate drivers based on HVES.

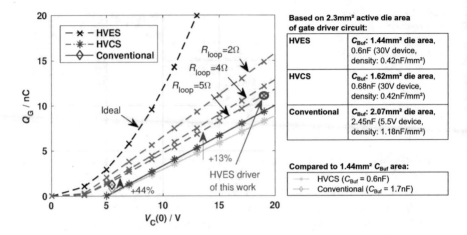

Fig. 4.23 Gate charge delivery versus precharge voltage level $V_C(0)$ for gate drivers based on various buffer capacitor implementation concepts, assuming the same size of the implemented circuits

When the quality factor of the gate loop is too small, near to 0.7 (see Sect. 2.1.4), the HVES circuit behaves similar to the HVCS concept (see Fig. 4.2). As HVCS has a lower circuit complexity, it might be preferable. Higher R_{par}, lower L_{loop}, and larger C_{loop} decrease the quality factor (see (2.1)). Hence, the following scenarios are attractive for the concept of HVCS:

- High switching speed is not acceptable for electromagnetic interference (EMI) issues in certain applications. HVES can accommodate slow-switching performance but requires a large inductor. However, the inductor might be too large for integration, or has a high parasitic resistance (more windings on the same chip area). A higher gate loop resistance R_{par} is also an option to slow down the gate drive speed (see also Sect. 2.1.4), but at the costs of lower quality factor. The HVCS concept allows an area-efficient capacitor integration even at large R_{par}.
- HVCS may be beneficial, if technology and available die area lead to a low quality factor, e.g., due to a high-voltage transistor in the gate loop that is only available with large specific on-resistance.
- The gate driver integrated on the same die as the power transistor typically has a negligibly small gate loop (see Sect. 2.1.4). Conventionally, the buffer capacitor is implemented as discrete component increasing the gate loop inductance again. The concept of HVCS allows for integrating the capacitor for medium Q_G at superior gate drive speed because of a negligible small gate loop inductance.

Figure 4.23 shows the gate charge delivery of the HVES circuit versus the pre-charge voltage $V_C(0)$ of drivers with different capacitor implementation methods. It assumes for all gate drivers the same total active die area. The implementation of the gate driver based on HVES in Fig. 4.9 occupies an active die area of 2.3 mm^2, comprising a capacitor with a size of 1.44 mm^2. Figure 4.23 shows how much gate charge can be delivered from gate drivers including all required circuitry, such as power transistors, active rectifier, etc., on a die area of 2.3 mm^2. In contrast, the diagram in Fig. 4.2 shows how much gate charge can be delivered from capacitors sized with 1.44 mm^2 of different capacitor implementation methods. This way, the comparison in Fig. 4.23 considers also the complexity of the gate driver circuits. For example, the conventional approach requires the smallest additional circuitry leading to a higher area portion for C_{Buf}. This results in 44 % more delivered gate charge compared to a buffer capacitor sized with 1.44 mm^2. The HVCS concept delivers 13 % more charge, compared to a capacitor occupying 1.44 mm^2. For the chosen parameter set and technology, the concept of HVES is still the most area efficient implementation. As shown in Fig. 4.23, the implemented gate driver based on HVES (Fig. 4.9) delivers up to 11.6 nC gate charge, which is about 10 times higher compared to a conventional gate driver (1.2 nC). It is sufficient to drive most commercially available GaN transistors.

4.4.3   Comparison with Prior Art

The comparison to prior art of integrated gate driver circuits is depicted in Table 4.2. The proposed driver is the only fully integrated gate driver that provides a three-level gate drive voltage (positive, 0 V, negative). Due to the implemented HVES circuits, it delivers up to 11.6 nC gate charge with a high gate drive current capability of 1.5 A without any external capacitors. This results in a gate charge delivery per buffer capacitor area of 4 nC mm^{-2}. Only [1], which is the driver presented in Fig. 3.9, achieves a higher value, since it also uses HVES, but it cannot support bipolar or three-level gate drive schemes. As expected, this also exceeds the value accomplished by the concept of HVCS in [3]. Besides, GaN transistors with an isolated gate, the driver supports a limited on-state gate current for GaN GITs. Nagai et al. [14], one of the early GaN driver publications, also supports GIT devices, but the gate current is very limited.

Table 4.2 Performance comparison with state of the art

Publication	This work	ISSCC'17 [1]	ISSCC'16 [12]	JSSC'15 [3]	ISSCC'15 [13]	ISSCC'12 [14]
Technology	Si, 0.18 µm BCD	Si, 0.18 µm BCD	Si, 0.35 µm BCD	Si, 0.18 µm BCD	Si, 0.35 µm BCD	GaN / Sapphire
Max. gate current	1.5 A	1.3 A	n.r.[a]	n.r.[a]	n.r.[a]	30 mA
Bipolar gate voltage	Yes	No	No	No	No	Yes
Three-level gate voltage	Yes	No	No	No	No	No
Support of GaN GIT	Yes	Yes	No	No	No	Yes
Buffer capacitors	3.6 nF (C_{DRV}), 2 × 0.6 nF (C_{HVx})	1.7 nF (5V), 0.6 nF (HV)	n.r.[a]	72.6 pF (5V), 18.4 pF (HV)	n.r.[a]	0
Max. gate charge	11.6 nC	11 nC	133 pC[c]	246 pC	133 pC[c]	n/a
Buffer capacitor area	2.9 mm^2	1.4 mm^2	0.718 mm$^{2[b]}$	0.065 mm^2	0.15 mm$^{2[b]}$	0
Q_G/buffer capacitor area	4 nC mm^{-2}	7.6 nC mm^{-2}	0.19 nC mm^{-2}	3.8 nC mm^{-2}	0.9 nC mm^{-2}	n/a
Max. V_{SW} slew rate	80 V/3 ns, 80 V/2.8 ns	n.r.[a]	40 V/1.5 ns, 40 V/1.2 ns	n.r.[a]	20 V/1 ns	n.r.[a]

[a] Not reported
[b] Extracted from chip micrograph
[c] Datasheet EPC8002

References

1. Seidel, A., & Wicht, B. (2017, February). 25.3 A 1.3A gate driver for GaN with fully integrated gate charge buffer capacitor delivering 11nC enabled by high-voltage energy storing. In *2017 IEEE International Solid-State Circuits Conference (ISSCC)* (pp. 432–433). https://doi.org/10.1109/ISSCC.2017.7870446

2. Seidel, A., & Wicht, B. (2018). Integrated gate drivers based on high-voltage energy storing for GaN transistors. *IEEE Journal of Solid-State Circuits*, 1–9. ISSN: 0018-9200. https://doi.org/10.1109/JSSC.2018.2866948

3. Seidel, A., Costa, M. S., Joos, J., & Wicht, B. (2015). Area efficient integrated gate drivers based on high-voltage charge storing. *IEEE Journal of Solid-State Circuits, 50*(7), 1550–1559. ISSN: 0018-9200. https://doi.org/10.1109/JSSC.2015.2410797

4. *GS66508T top-side cooled 650 V E-mode GaN transistor, preliminary datasheet* (2018). GaN Systems Incorporation.

5. Qin, S., Lei, Y., Barth, C., Liu, W., & Pilawa-Podgurski, R. C. N. (2015, September). A high-efficiency high energy density buffer architecture for power pulsation decoupling in grid-interfaced converters. In *Proceedings of IEEE Energy Conversion Congress and Exposition (ECCE)* (pp. 149–157). https://doi.org/10.1109/ECCE.2015.7309682

6. Kaufmann, M., Seidel, A., & Wicht, B. (2020, March). Long, short, monolithic – The gate loop challenge for GaN drivers: Invited paper. In *Proceedings of IEEE Custom Integrated Circuits Conference (CICC)* (pp. 1–5). https://doi.org/10.1109/CICC48029.2020.9075937

7. Moench, S., Hillenbrand, P., Hengel, P., & Kallfass, I. (2017, October). Pulsed measurement of sub-nanosecond 1000 V/ns switching 600 V GaN HEMTs using 1.5 GHz low-impedance voltage probe and 50 ohm scope. In *Proceedings of IEEE 5th Workshop Wide Bandgap Power Devices and Applications (WiPDA)* (pp. 132–137). https://doi.org/10.1109/WiPDA.2017.8170535

8. Schuylenbergh, K., & van and Puers, R. (2009). *Inductive powering – Basic theory and application to biomedical systems*. Berlin Heidelberg: Springer Science & Business Media. ISBN: 978-9-048-12412-1.

9. Bau, P., Cousineau, M., Cougo, B., Richardeau, F., Colin, D., & Rouger, N. (2018, July). A CMOS gate driver with ultra-fast dV/dt embedded control dedicated to optimum EMI and turn-on losses management for GaN power transistors. In *Proceedings of 14th Conference on Ph.D. Research in Microelectronics and Electronics (PRIME)* (pp. 105–108). https://doi.org/10.1109/PRIME.2018.8430331

10. de Vries, I. D. (2002, March). A resonant power MOSFET/IGBT gate driver. In *Proceedings of APEC Seventeenth Annual IEEE Applied Power Electronics Conference and Exposition (Cat. No.02CH37335)* (Vol. 1, pp. 179–185). https://doi.org/10.1109/APEC.2002.989245

11. Perez, A., Jorda, X., Godignon, P., Vellvehi, M., Galvez, J. L., & Millan, J. (2003, September). An IGBT gate driver integrated circuit with full-bridge output stage and short circuit protections. In *Semiconductor Conference, 2003. CAS 2003. International* (Vol. 2, p. 248). https://doi.org/10.1109/SMICND.2003.1252427

12. Ke, X., Sankman, J., Song, M. K., Forghani, P., & Ma, D. B. (2016, January). 16.8 A 3-to-40V 10-to-30MHz automotive-use GaN driver with active BST balancing and VSW dual-edge dead-time modulation achieving 8.3% efficiency improvement and 3.4ns constant propagation delay. In *Proceedings of IEEE International Solid-State Circuits Conference (ISSCC)* (pp. 302–304). https://doi.org/10.1109/ISSCC.2016.7418027

13. Song, M. K., Chen, L., Sankman, J., Terry, S., & Ma, D. (2015, February). 16.7 A 20V 8.4W 20MHz four-phase GaN DC-DC converter with fully on-chip dual-SR bootstrapped GaN FET driver achieving 4ns constant propagation delay and 1ns switching rise time. In *2015 IEEE International Solid-State Circuits Conference (ISSCC) Digest of Technical Papers* (pp. 1–3). https://doi.org/10.1109/ISSCC.2015.7063046

14. Nagai, S., Negoro, N., Fukuda, T., Otsuka, N., Sakai, H., Ueda, T., et al. (2012, February). A DC-isolated gate drive IC with drive-by-microwave technology for power switching devices. In *2012 IEEE International Solid-State Circuits Conference* (pp. 404–406). https://doi.org/10.1109/ISSCC.2012.6177066

Chapter 5
Gate Drivers for Large Gate Loops Based on HVES

5.1 Introduction

In many applications, large distances between the driver and the power transistor are required or offer advantages, because the gate driver and power transistor are physically separated. Extreme conditions, like higher temperatures at the gallium nitride (GaN) transistor due to losses, affect only the power transistor and not the gate driver. This is especially true, because harsh environments are a promising field for GaN transistors [1–3]. Different optimized assemblies for the power transistor and the driver design help to exploit the switching and load current capability of the power transistor. For example, a multilayer substrate for the gate driver enables compact wiring, while a power transistor on a ceramic substrate offers better performance for high currents and temperatures [4]. A power transistor board without gate drivers enables more effective cooling, e.g., by placing heat sinks tightly on both sides of the substrate [5]. In some cases, large gate loops occur due to given distances and tight space constraints in a power module or an engine [5].

In any case, typical gate driver designs are optimized to be as close as possible to the power transistor to form a small gate loop that results in a small gate loop inductance. According to Sect. 4.1.2, a small gate loop inductance significantly increases the gate drive speed (see also Sect. 2.1.4), which is crucial for low power-transistor switching losses (Sect. 2.1.8). Section 2.3 shows the importance of a small gate loop inductance to keep the gate voltage within a specified range to be robust against interferences like Miller coupling. An additional capacitor in parallel to the gate-source terminals of the power transistor buffers these interferences, but it increases the overall gate capacitance, further slows down the gate drive speed, and adds to the gate drive losses [6]. These drawbacks can be addressed utilizing the concept of high-voltage energy storing (HVES) [7, 8].

© The Author(s), under exclusive license to Springer Nature Switzerland AG 2021
A. Seidel, B. Wicht, *Highly Integrated Gate Drivers for Si and GaN Power Transistors*, https://doi.org/10.1007/978-3-030-68940-7_5

5.2   Concept

As shown in Sect. 4.1.2, HVES is superior compared to other gate drive concepts in terms of gate drive speed assuming the same gate loop inductance.

While in Chap. 4 additional integrated inductors are implemented in the gate loop, the gate driver presented below is based on HVES that utilizes the parasitic gate loop inductance as resonance inductor. The driver enables a fast and robust gate drive operation even at larger distances between the gate driver and the power transistor resulting in larger loop inductance L_{loop}, and added parasitic gate-source capacitance $C_{GS,add}$. According to Sects. 2.1.4 and 4.1.2, the conventional gate driver has some gate overshoot, caused by the remaining energy in the gate loop inductance L_{loop} when the gate is already fully charged. This requires typically a damping resistor R_G to damp the gate overshoot. Gate drivers based on HVES instead utilize the energy of that inductance to charge the gate.

Figure 5.1 derives the proposed gate driver concept. For comparison, Fig. 5.1a shows the concept of the gate driver based on HVES presented in Chap. 4. Figure 5.1b demonstrates the basic concept of the gate driver for large gate loops. Both concepts utilize HVES and provide a bipolar gate drive scheme. For turning on and off the gate driver, energy oscillates resonantly in the gate loop to charge and discharge C_G to a positive and negative gate voltage, respectively. The energy E_p and E_n is added during every driver turn-on and turn-off event compensating the energy losses in the gate loop due to parasitic resistances. The HVES circuit in Fig. 5.1a comprises the inductors L_{HVp}/L_{HVn}, the high-voltage capacitors indicated as energy delivering elements E_p, and E_n and the diodes D_{Rp} and D_{Rn}.

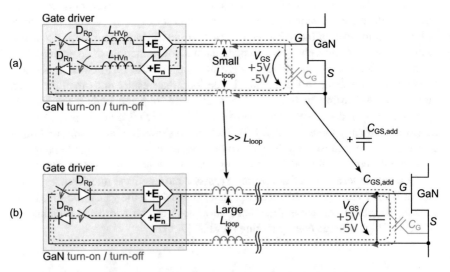

Fig. 5.1 (**a**) The concept of the HVES-based gate driver and (**b**) proposed concept of the gate driver for large gate loops (Chap. 4)

While the gate driver in Fig. 5.1a is optimized for a full integration of the buffer capacitors, the proposed gate driver is optimized for high-performance switching coping with large gate loops (L_{loop}). The driver in Fig. 5.1b requires no on-chip inductors L_{HVp} and L_{HVn} anymore, because the larger parasitic gate loop inductance L_{loop} replaces this function. A discrete capacitor $C_{GS,add}$, in parallel to the gate capacitance C_G, buffers interferences like Miller coupling (see Sect. 2.3). For this purpose, the implemented gate drivers in Figs. 4.10 and 4.11 of the concept in Fig. 5.1a have an integrated low-voltage buffer capacitor C_{DRV} that needs to be close to the GaN transistor. Because of the additional capacitive driver load $C_{GS,add}$ for the gate driver in Fig. 5.1b, the high-voltage buffer capacitors of the HVES circuits need to be larger, which makes their integration more difficult. This concept is similar to the resonant gate driver in [9], which has also a large gate loop capability but with the drawback of lower gate drive speed since it is not based on the concept of HVES (see Sect. 4.2).

5.3   Implementation

Figure 5.2 shows the implementation of the proposed gate driver for large gate loops. Figure 5.3 depicts its timing. At turn-on, the high-voltage transistors M_{Hp} and M_{Ln} trigger a resonance pulse, charging the gate of the power transistor and $C_{GS,add}$ from the high-voltage capacitor C_{HVp}. The free-wheeling diode D_{FWp} conducts the current as long as the parasitic gate loop inductance L_{loop} draws the current. The rectifier diode D_{Rp} prevents a charge backflow from the gate. Simultaneously to the switching of M_{Hp}, MN_p turns on keeping the gate to V_{DRV} via a high-ohmic resistor R_{Bp}. Charge flows via R_{Bp} into or out of the gate in case that the HVES

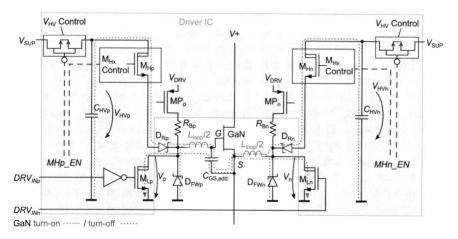

Fig. 5.2 Implementation of gate driver for large gate loops

Fig. 5.3 Waveforms of the
gate driver for large gate
loops of the circuit in Fig. 5.2

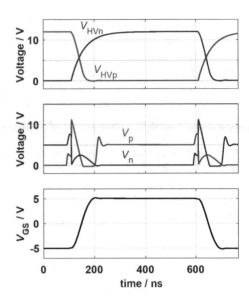

circuit delivers not exactly the required amount of gate charge. The sizing of R_{Bp} is considered further below in this section. In contrast to the gate driver in Chap. 4, the required gate charge varies only slightly since it is mainly determined by the capacitor $C_{GS,add}$ and not by C_G. Furthermore, R_{Bp} limits the current into V_{DRV} when the high-voltage V_{HVp} is applied at the gate driver output during the turn-on process. This high voltage at the driver output occurs because V_{HVp} drops mainly across the gate loop inductance outside the gate driver, in contrast to the gate drivers in Chap. 4 with integrated inductors L_{HVp} and L_{HVn}. The same procedure takes place at driver turn-off at the GaN transistor's source side, charging the source via M_{Hn}, D_{Rn}, and M_{Lp} to $V_{GS} = -V_{DRV}$.

The block "M_{Hx} Control" is similar to the "M_{HVx} Control" block in Fig. 4.16. Since in Fig. 5.2 the sources of M_{Hp}/M_{Hn} achieve voltage levels significantly above V_{DRV}, a high-voltage level shifter is implemented.

If implemented at the high-side, the gate driver output nodes V_p and V_n need to sink and source common mode currents, as described in Sect. 4.2.3. However, the gate driver output nodes V_p and V_n cannot sink the current when M_{Lp} or M_{Ln} are not conducting. Therefore, voltage levels at the nodes V_p/V_n may exceed the maximum ratings, which can be prevented by clamping diodes connected between V_{SUP} and V_p, and V_{SUP} and V_n. The operation as high-side gate driver needs to be implemented carefully to minimize parasitic capacitances to the application's low side and, thus, to keep common mode currents low (see Sect. 4.2.3). Due to the large gate loop inductance, oscillations may occur between the gate and this parasitic capacitance. This can be damped by a common mode choke, implemented in the driver supply path (see Fig. 5.7 in Sect. 5.4) or, if necessary, also in the signal transmission path. The gate driver, applied to a high-side transistor or in environments causing high common mode currents, is not yet fully experimentally validated and is worth for further evaluation.

Sizing of Capacitor $C_{GS,add}$ The capacitor $C_{GS,add}$ (Fig. 5.2) buffers the gate source voltage V_{GS} of the power transistor during interferences, like the Miller coupling (see Sect. 2.3.1). While the driver is in off-stage, the Miller coupling lifts V_{GS} towards the threshold voltage V_{th} of the power transistor by ΔV_{GS}. To achieve a sufficient margin, the sizing of $C_{GS,add}$ complies with the amount of Miller charge Q_{Miller} coupling into the gate and the V_{GS} applied during driver off-state. ΔV_{GS} can be calculated from

$$\Delta V_{GS} = \frac{Q_{Miller}}{C_{GS,add} + C_G}, \tag{5.1}$$

while $C_{GS,add}+C_G$ is the total capacitance at the gate-source terminals of the power transistor. Rearranging (5.1), leads to

$$C_{GS,add} = \frac{Q_{Miller}}{\Delta V_{GS}} - C_G, \tag{5.2}$$

For example, the GaN transistor PGA26E07BA (Panasonic) is specified with $C_G= 405\,pF$ and $Q_{Miller}= 2.6\,nC$ (Q_{Miller} corresponds to Q_{GD} value in datasheet) [10]. Based on (5.2), $C_{GS,add}$ results in 2.2 nF assuming $\Delta V_{GS}= 1\,V$.

Gate Overshoot and Oscillations Section 4.1.2 describes that the output capacitance of the gate driver causes gate over- and undershoots, or even ringing. A gate charge backflow from C_G into C_{Buf} again or an oscillating between these capacitances occur at the end of a driver switching event. This can be also caused by the rectifier diodes/active rectifiers D_{Rp}, D_{Rn} (Fig. 5.2) that have reverse recovery behavior. It applies to all gate drivers based on HVES as described in Sect. 4.4.1, but it has more effect in case of large gate loops inductance L_{loop} (see Sect. 2.1.4). The measurements at the gate driver in Chap. 4 show almost no gate overshoot (Fig. 4.20) as the driver has a very small L_{loop} of ~6 nH between the gate driver output and the gate of the GaN transistor (Fig. 4.19). Furthermore, due to the different driver circuit topology, the high-voltage blocking freewheeling diodes D_{FWp} and D_{FWn} in Fig. 4.11 are not adding to the gate driver output capacitance. The active rectifiers D_{Rp} and D_{Rn} with their detailed circuit in Fig. 4.16 have a blocking capability of only 5 V resulting in a small component size and thus smaller parasitic capacitances and reverse recovery. However, the gate driver circuits in this chapter are tested with L_{loop} of about 600 nH. Thus, the rectifier diodes D_{Rp} and D_{Rn} and the free-wheeling diodes D_{FWp} and D_{FWn} are implemented as discrete Schottky diodes of type BAS85 [11] with very low parasitic junction capacitances and a blocking voltage up to 30 V. Schottky diodes exhibit similar or even lower junction capacitances with increasing blocking voltage ratings [12–14]. A high current capability of the diodes increases their junction capacitances, which is necessary when operating at very high gate charge speed. In case that the gate overshoot is too high, Fig. 5.4 shows

Fig. 5.4 Modified circuit
implementation of the gate
driver output stage with
minimized parasitic output
capacitance

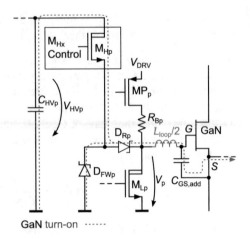

GaN turn-on ······

a modification of the gate driver circuit in Fig. 5.2. The modification is depicted
exemplarily for the driver circuit part at the gate of the GaN transistor. The same
needs to be done for the circuit at the source side. In contrast to Fig. 5.2, the free-
wheeling diode D_{FWp} is implemented before the rectifier diode D_{Rp} reducing the
gate driver output capacitance and thus the gate overshoot. It should be noticed that
with this circuit modification, more losses occur during free-wheeling because the
current flows additionally through D_{Rp} causing a voltage drop.

A further action against oscillations in the gate loop is to reduce R_{Bp} and R_{Bn}.
R_{Bp} and R_{Bn} are implemented to constantly keep the gate driver output nodes V_p
and V_n at V_{DRV} level during GaN transistor on-state and off-state, respectively. In
case of oscillations, the remaining energy from L_{loop} flows back from the transistor
causing an overshoot at the driver output nodes V_p/V_n, shortly after a GaN transistor
turn-on/turn-off (see Fig. 5.3). Lower values for R_{Bp} and R_{Bn} reduce this overshoot
by dissipating the energy of that oscillation. A trade-off has to be found for R_{Bp}
and R_{Bn}, because smaller resistor values conduct more current to V_{DRV} during GaN
turn-on/turn-off process. This makes larger capacitors C_{HVp}/C_{HVn} necessary and
increases the driver power consumption.

Nevertheless, the experimental results in Sect. 5.4 show very low gate overshoots
without circuit modification for a distance between gate driver and power transistor
of ~25 cm at an adequate gate charge speed.

5.4 Experimental Results and Comparison to Prior Art

The gate driver for large gate loops has been implemented in a 0.18 μm BCD
technology. This first implementation consists of two dies, one for the gate and one
for the source side of the GaN transistor. This can be also realized on a single die.

The capacitors C_{HVp} and C_{HVn} are 2.2 nF discrete devices. In case of smaller power transistors, these could be integrated.

In order to evaluate the expected behavior in case of a large gate connection, the proposed gate driver (Fig. 5.2) is measured in comparison to a conventional gate driver (ADuM3223). Figure 5.5a shows the measurement setup and Fig. 5.6a the corresponding waveforms. Both gate drivers provide bipolar gate voltages. The architecture of the conventional driver corresponds to the circuit in Fig. 2.13. With a gate resistor of $R_G = 15\,\Omega$, the V_{GS} of the conventional gate driver has a gate overshoot of ~0.5 V and a rise/fall time of 80 ns. With a similar gate overshoot, the proposed gate driver (without R_G) achieves a rise and fall time of 47 ns, which is an improvement of ~40 %. The conventional driver with a gate resistor R_G of only 5 Ω leads to a similar rise and fall time but results in a significant gate overshoot and ringing (of up to ~2.5 V).

The gate overshoot is a result of the remaining energy in the gate loop inductor at the end of the gate charge event (see Sect. 2.1.4). A clamping circuit is a possible solution dissipating this energy. The clamping circuit, Fig. 5.5b, comprises two zener diodes clamping the negative and positive V_{GS}, respectively. The Schottky diodes in series are optional to minimize the effect of reverse recovery charges of the zener diodes. With an R_G of only 5 Ω, the results in Fig. 5.6b show fast V_{GS} rise and fall times of 54 ns, which is still ~15 % more than that of the gate driver based on HVES. For the proposed gate driver circuit, conceptually no oscillations occur (see Sect. 5.2). Oscillations caused by non-ideal components are damped due to small gate driver output capacitances and the resistors R_{Bp}/R_{Bn} (see Fig. 5.2), as described in Sect. 5.3. Finally, a measurement without any gate resistor ($R_G = 0\,\Omega$) was performed at the conventional gate driver with clamping circuit. As expected, the V_{GS} waveform exhibits ringing and only slightly shorter rise and fall times of 45 ns, which is not shown in Fig. 5.6b for clarity reasons. While this represents a limit of the conventional approach, the proposed concept has the potential for even faster switching, as described in Sect. 4.1.2. Moreover, another conceptual gate driver implementation option with theoretically faster gate drive capability is presented in Sect. 6.3.

The proposed gate driver falls into the category of resonant gate driver concepts (see Sect. 2.1.6). It is a fast switching resonant gate driver that utilizes the energy in the gate loop inductance for charging the gate. This provides the potential of higher gate driver efficiency (see Sect. 4.1.3). Conventional resonant approaches require a clamping circuit near to the power transistor that feeds back into the gate driver

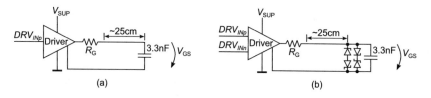

Fig. 5.5 Gate driver configurations (**a**) without and (**b**) with clamping structure

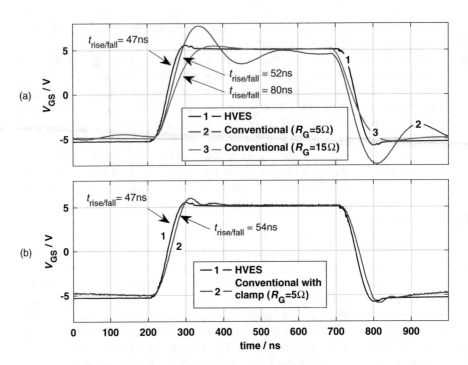

Fig. 5.6 Measured V_{GS} of the gate driver for large gate loop, compared to conventional gate driver with setups (**a**) and (**b**), according to Fig. 5.5a and b, respectively

supply to save energy. Typically, this is realized by a diode or transistor connected from the gate of the power transistor to the driver supply, like in [15, 16]. Therefore, they are not suitable for large gate loops. The resonant gate drivers in [9] and [17] have a large gate loop capability but with the drawback of lower gate drive speed, since they are not based on the concept of HVES. Furthermore, they do not have a buffer capacitor in parallel to the gate source terminals of the power transistor, and are therefore less robust against Miller coupling (see Sect. 2.3.1).

Figure 5.7 shows the evaluation of the gate driver in a double pulse test setup. An arbitrary waveform generator controls the driver via a galvanically isolated gate driver SI8275GB-IS1, which has only 0.5 pF coupling capacitance between the primary and the secondary isolation side. As gate driver power supply, the DCDC converter NMS0509C is well suitable for fast switching applications with a coupling capacitance of only 1.9 pF. Low coupling capacitance between the control and the power transistor side is crucial for low common mode coupling currents (see Sects. 2.4.5 and 5.3). The common mode choke CMC (type 0805USB-172ML from Coilcraft) further reduces the common mode current through the power supply path. The gate driver works with a ∼50 cm gate loop (2× 25 cm) with the power device at $V_+ = 400\,\text{V}$ and a load current of $I_L = 10\,\text{A}$. A GaN gate injection transistor (GIT) (PGA26E19BA, Panasonic) is chosen for the test. Very fast voltage transitions are performed, ∼60 V ns^{-1} for GaN transistor turn-on and ∼40 V ns^{-1}

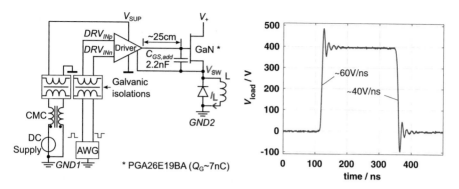

Fig. 5.7 Transient load voltage measurement of the gate driver for large gate loop in a double pulse test setup

for turn-off, respectively. The V_{GS} waveform was not measured, because a voltage probe at the terminals of the GaN transistor would affect the measurement results. Because a GIT clamps the gate-source voltage V_{GS} only in positive direction (see Sect. 2.2.3), the proposed gate driver has the advantage of providing a low undershoot for the negative V_{GS}, in addition to the gate drive speed advantage. Overall, the experimental results confirm a superior switching performance of the proposed gate driver, with high gate drive speed and low gate overshoot.

References

1. Hassan, A., Savaria, Y., & Sawan, M. (2018). Electronics and packaging intended for emerging harsh environment applications: a review. *IEEE Transactions on Very Large Scale Integration (VLSI) Systems, 26*(10), 2085–2098. ISSN: 1063-8210. https://doi.org/10.1109/TVLSI.2018.2834499

2. Hassan, A., Savaria, Y., & Sawan, M. (2018). GaN integration technology an ideal candidate for high-temperature applications: A review. *IEEE Access*, 1. ISSN: 2169-3536. https://doi.org/10.1109/ACCESS.2018.2885285

3. Moroney, M. (2015). *New power switch technology and the changing landscape for isolated gate drivers*. Norwood: Analog Devices, Inc.

4. Yu, C., Buttay, C., & Labouré, É. (2017). Thermal management and electromagnetic analysis for GaN devices packaging on DBC substrate. *IEEE Transactions on Power Electronics, 32*(2), 906–910. ISSN: 0885-8993. https://doi.org/10.1109/TPEL.2016.2585658

5. März, M., Schletz, A., Eckardt, B., Egelkraut, S., & Rauh, H. (2010, March). Power electronics system integration for electric and hybrid vehicles. In *Proceedings of 6th International Conference on Integrated Power Electronics Systems* (pp. 1–10)

6. *How to drive GaN enhancement mode power switching transistors* (2014, October). GaN Systems.

7. Kaufmann, M., Lueders, M., Kaya, C., & Wicht, B. (2020). 18.2 A monolithic E-mode GaN 15W 400V offline self-supplied hysteretic buck converter with 95.6% efficiency. In *Proceedings of IEEE International Solid- State Circuits Conference – (ISSCC)* (pp. 288–290).

8. Seidel, A., & Wicht, B. (2018). Integrated gate drivers based on high-voltage energy storing for GaN transistors. *IEEE Journal of Solid-State Circuits*, 1–9. ISSN: 0018-9200. https://doi.org/10.1109/JSSC.2018.2866948

9. de Vries, I. D. (2002, March). A resonant power MOSFET/IGBT gate driver. In *Proceedings of APEC Seventeenth Annual IEEE Applied Power Electronics Conference and Exposition (Cat. No.02CH37335)* (Vol. 1, pp. 179–185). https://doi.org/10.1109/APEC.2002.989245

10. *PGA26E07BA Preliminary Datasheet* (2016, October). Panasonic Corporation.

11. *BAS85 – Small Signal Schottky Diode* (2017, June). Vishay Intertechnology Inc.

12. *RB541VM-40 – Schottky Barrier Diode* (2016, September). ROHM Co.

13. *RB541VM-30 – Schottky Barrier Diode* (2016, September). ROHM Co.

14. *RB558VYM150FH – Schottky Barrier Diode* (2017, April). ROHM Co.

15. Eberle, W., Zhang, Z., Liu, Y., & Sen, P. C. (2008). A current source gate driver achieving switching loss savings and gate energy recovery at 1-MHz. *IEEE Transactions on Power Electronics, 23*(2), 678–691. ISSN: 0885-8993. https://doi.org/10.1109/TPEL.2007.915769

16. Long, Y., Zhang, W., Costinett, D., Blalock, B. B., & Jenkins, L. L. (2015, March). A high-frequency resonant gate driver for enhancement-mode GaN power devices. In *Applied Power Electronics Conference and Exposition (APEC), 2015 IEEE* (pp. 1961–1965). https://doi.org/10.1109/APEC.2015.7104616

17. Fujita, H. (2010, April). A resonant gate-drive circuit capable of high-frequency and high-efficiency operation. *IEEE Transactions on Power Electronics, 25*(4), 962–969. ISSN: 0885-8993. https://doi.org/10.1109/TPEL.2009.2030201

Chapter 6
Outlook and Future Work

6.1 Gate Drivers Based on High-Voltage Charge Storing (HVCS)

The gate driver output stage circuits based on high-voltage charge storing (HVCS) [1] in Chap. 3 turned out to be very area efficient. Further improvement applies to the circuit options in Figs. 3.1 and 3.3. An additional charging path from V_3 to the gate of MN_1 via a high-voltage NMOS transistor could deliver considerable gate charge. The additional path charges the gate until V_{B1} rises above V_3. This enables smaller bootstrap capacitors, further reducing the area of the gate driver output stage. This may require modifications to optimize the timing. A similar additional current path is also implemented in [2], which achieves a very small gate driver output stage, but requires PMOS transistors and complex control.

Gate drivers based on HVCS can be used with other transistor types, such as discrete silicon or SiC transistors that require gate charges in a suitable range for capacitor integration.

The monolithic integration of gate drivers on the same die with the power transistor is a growing market. The concept of HVCS can be applied to achieve the integration of the buffer capacitor as well (see Sect. 4.4.2). Implementing smart supporting circuits, like the implementation of very fast over-current protection, over-temperature protection, or slope shaping, might give interesting research topics [3].

6.2 Gate Drivers Based on High-Voltage Energy Storing (HVES)

The gate drivers based on HVES [4] can be extended to allow an active gate control (see Sect. 2.3.2) by working with parallel gate charge paths as depicted in Fig. 6.1.

© The Author(s), under exclusive license to Springer Nature Switzerland AG 2021
A. Seidel, B. Wicht, *Highly Integrated Gate Drivers for Si and GaN Power Transistors*, https://doi.org/10.1007/978-3-030-68940-7_6

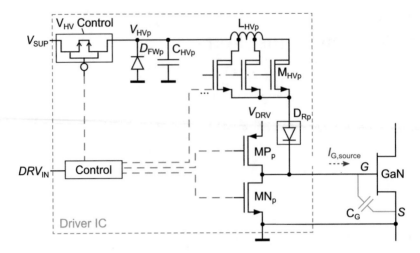

Fig. 6.1 Gate driver concept with active gate control by parallel gate charge paths from an HVES circuit

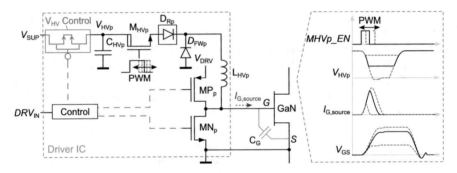

Fig. 6.2 Gate voltage control by adapting the pulse-width of M_{HVp}

While Fig. 6.1 shows a unipolar gate drive scheme, the concept can be applied also to the source side of the gallium nitride (GaN) transistor. The gate current flows over several intermediate coil taps that lead to different gate speed behavior. The shaping of the switching transition may provide benefits, such as an optimization of losses and EMC (see Sect. 2.3.2).

Another extension of the concept addresses the control of the gate voltage level. As shown in Fig. 6.2, such a gate voltage control can be implemented in the HVES circuit by changing the turn-on pulse width of the high-voltage transistor M_{HVp}. This is achieved by rearranging the inductor L_{HVp} and the freewheeling diode D_{FWp} compared to the circuit in Fig. 4.9.

In Fig. 6.2, the degree to which the capacitor C_{HVp} is discharged specifies the energy portion transferred to the gate. Optionally, this approach allows for a constant voltage source at C_{HVp} (node V_{HVp}), because the on-time of M_{HVp} determines the amount of energy transferred to the gate. In this case, the "V_{HV} Control" block is not needed anymore. The high voltage at node V_{HVp} leads to a fast gate charge. Furthermore, a constant voltage source at node V_{HVp} avoids losses caused by recharging C_{HVp} once per switching cycle, increasing the efficiency of the circuit (see Sect. 4.1.3). However , the buffer capacitor buffering V_{HVp} needs to be large and it is typically a discrete component.

The gate drivers based on HVES of this work operate in buck mode, i.e., it converts $V_C(0) > 5\,\text{V}$ to $V_{GS} = 5\,\text{V}$. The concept has superior Q_G-delivery capability in the boost mode ($V_C(0) < 5\,\text{V}$) as well (see Fig. 4.2). The boost mode can be interesting for low-voltage driver designs to achieve gate drive voltages higher than the gate driver supply voltage. This is similar to the resonant gate drivers in [5–8], which support a gate drive voltage boosting.

Besides GaN transistors, drivers based on HVES can be applied to other transistor types, such as silicon or SiC transistors, that exhibit gate charges in a range that allows for buffer capacitor integration.

As the concept of HVES is a general solution for a very area-efficient buffer capacitor integration that delivers pulsed currents, it may also be suitable for other applications, like for highly integrated charge pump circuits.

6.3   Gate Drivers for Large Gate Loops

The gate driver for large gate loops [9] was tested in double pulse measurement as high-side transistor. However, the gate-source voltage of the GaN transistor could not be measured due to limits set by the measurement equipment. The gate driver requires further investigations in terms of robustness and common-mode transient immunity (CMTI). Furthermore, EMI measurements should be performed.

The gate driver for large gate loops can be applied to other transistor types, like silicon carbide (SiC).

A second implementation option of the gate driver can be investigated in more detail and implemented on chip for experimental evaluation.

Figure 6.3 shows a conceptual circuit of the second gate driver option for large gate loops, with the corresponding waveforms in Fig. 6.4. Compared to the implementation in Fig. 5.2, it is also based on two HVES circuits that deliver charge onto the gate and the source of the GaN transistor, respectively, to obtain a bipolar gate drive scheme. The basic differences, i.e., advantages and disadvantages, are discussed further below in this section.

For turning on the GaN transistor, M_{Hp} and the buffer B_p turn on. C_{HVp} discharges resonantly via the gate loop inductance charging the gate to $V_{GS} = V_{DRV}$ and charging C_{HVn} via the rectifier diode D_{Rp} and the inverter I_n. If too much or too little charge comes from C_{HVp}, the buffer B_p pulls the gate to V_{DRV} over R_{Bp}, and keeps it at this level. Accordingly, to turn off the GaN transistor, C_{HVn} discharges and charges V_{GS} to $-V_{DRV}$, via C_{HVp} and D_{Rn}. Hence, the energy oscillates between C_{HVp} and C_{HVn} when turning on or off the gate driver. The buffer B_p and inverter I_n add energy from V_{DRV} during GaN transistor turn-on and turn-off, respectively, to

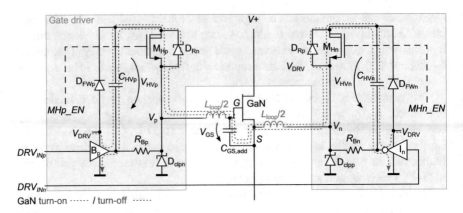

Fig. 6.3 Option of the gate driver for large gate loops

Fig. 6.4 Waveforms of the gate driver for large gate loops of the circuit in Fig. 6.3

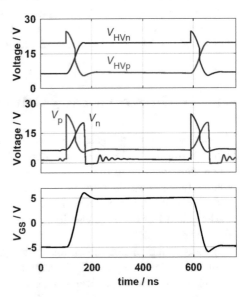

compensate the losses in the gate loop. The addition of energy from a low-voltage supply to a resonance circuit that oscillates at higher voltage than the supply voltage is well known in class D oscillator circuits that are used, for example, in an inductive link [10]. For a gate voltage regulation, short pulses can be applied at the output of the buffer B_p and the inverter I_n, respectively, triggering the turn-on and turn-off events. The duration of the pulses adapts the energy added to the resonance circuit, i.e., the gate loop, every switching cycle. R_{Bp}/R_{Bn} need to be controlled separately. The diodes D_{clpp}/D_{clpn} prevent the voltage nodes V_p and V_n from falling below ground shortly after the GaN transistor gate is charged/discharged (see Fig. 6.4).

The advantages and disadvantages of this circuit compared to Fig. 5.2 are as follows.

Advantages:

- Low-voltage driver supply possible (e.g., 5 V), without giving up the benefits of a HVES, and consequently the high gate drive speed.
- An even higher gate drive speed can be performed, because the high-voltage capacitors C_{HVp}/ C_{HVn} take effect during the whole gate charge/discharge process. This means, they are not short-circuited via the freewheeling diodes D_{FWp}/D_{FWn} (see Sect. 4.1.2). The freewheeling diodes are not necessary in ideal case (only for preventing an asymmetric C_{HVp}/ C_{HVn} discharge due to tolerances in real design). The speed advantage is particularly effective with high initial voltages V_{HVp}/V_{HVn}.
- Better efficiency, since this approach avoids the recharge losses of C_{HVx} described in Sect. 4.1.3.
- Only one high-voltage transistor in the current path leads to a higher quality factor of the gate loop resonance circuit, or smaller chip area.
- Compared to Fig. 5.2, the "V_{HV} Control" blocks for recharging C_{HVp}/C_{HVn} are not required anymore.

Disadvantages:

- Since the capacitors C_{HVp} and C_{HVn} are connected in series during charging/discharging the gate, they must be twice in size to achieve the same total capacitance value as the implementation option in Fig. 5.2. Furthermore, C_{HVp} and C_{HVn} need to be dimensioned larger with more energy storage capability because at the end of a switching event not only the gate capacitance is charged but also partially the opposite high-voltage capacitor C_{HVp} or C_{HVn}. Hence, the original implementation option in Fig. 5.2 is more suitable in case of aiming an area-efficient capacitor integration.
- An additional circuit (e.g. a charge pump) is required to provide the initial voltage at C_{HVp} (e.g. 40 V) at start-up.
- The concept tends to have a higher gate over- and undershoot (see Fig. 6.4). This is because the driver outputs V_p/V_n have a higher total output capacitance caused by the high-voltage transistors M_{Hp}/M_{Hn}, the rectifier diodes D_{Rp}/D_{Rn}, and the diodes D_{clpp}/D_{clpn}. The gate over- and undershoot can be significantly reduced by replacing the diodes D_{clpp}/D_{clpn} with small high-voltage transistors actively clamping the nodes to chip ground with some delay after turn-on/-off. These transistors dissipate the energy stored in the gate driver output nodes V_p/V_n before discharging the gate capacitance of the GaN transistor. Since this clamping needs accurate timing, a possible approach might be to automatically trigger if V_p/V_n exceed a specified voltage within a time window shortly after a GaN transistor turn-on/-off event.
- This concept might be more sensitive for operation as high-side gate driver, because common mode currents can charge the capacitors C_{HVp}/C_{HVn} leading to a varying amount of energy transferred to the gate. This possible issue requires some further investigations, and preventive actions may need to be installed.

The simulation results in Fig. 6.4 show very fast V_{GS} rise and fall times of ∼35 ns. The gate over- and undershoot of ∼1.2 V can be significantly reduced by replacing the diodes D_{clpp}/D_{clpn} by an active clamping as just described.

6.4 Boost Converter High-Side Driver Supply

Section 2.4.2 shows significant advantages of a bootstrap circuit as a high-side gate driver supply, mainly regarding compactness and low complexity. However, it also has some limitations, e.g., the minimum and maximum allowed duty cycle, or the driver supply possibility during application start-up.

Figure 6.5 shows a proposed high-side driver supply via conventional boost converter comprising L_1, D_1, and MN_1. Without the components L_1 and MN_1, it would be a conventional bootstrap supply (see Sect. 2.4.2). MN_1 needs to be rated for relatively low currents, but it must withstand high voltage levels of ∼V_+. A PWM signal controls MN_1, and the duty cycle determines how strongly C_{BST} is charged. Since there is no voltage feedback from C_{BST}, a linear regulator (LDO) or a zener diode on the high-side can be used for regulation. Such components are often implemented anyway in a conventional bootstrap circuit to prevent overvoltages at the bootstrap capacitor (see also Sect. 2.4.2). In contrast to a bootstrap circuit, the boost converter can charge C_{BST} also when the low-side power transistor is off via the half-bridge high-side power transistor. This enables to supply a 100 % turned-on high-side power transistor, as well as a safe supply during application start-up. If the low-side power transistor is in on-state, C_{BST} will be recharged similarly to a conventional bootstrap circuit. The additional inductance L_1 in the charge path prevents current peaks [11].

If C_{BST} is not implemented, this driver supply has a current source characteristic by L_1 charging C_{DRV} resonantly. Nevertheless, a fast linear regulator (LDO) or a zener diode can be installed to limit the voltage at C_{DRV}. This may allow a more efficient recharging which takes significant effect for small C_{BST} (see Sect. 4.1.3).

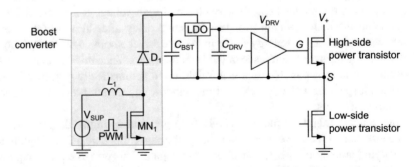

Fig. 6.5 Proposed high-side gate driver supply via boost converter

However, more detailed loss considerations need to be done in order to be able to make a final statement in this point.

A low parasitic coupling capacitance between high-side and low-side is crucial to achieve a high CMTI (see Sect. 2.4.5). As L_1 blocks transient currents, the effective coupling capacitance is formed from the junction capacitance of D_1 and the series-connected drain-source capacitance of MN_1. This is lower compared to a conventional bootstrap circuit, where only the junction capacitance of D_1 takes effect. This points to the fact that the proposed driver supply circuit may have a higher CMTI.

This circuit provides a low-cost, compact gate driver power supply that overcomes some limitations of a conventional bootstrap supply. It is not yet evaluated by simulation or experimental results and requires further investigations.

References

1. Seidel, A., Costa, M. S., Joos, J., & Wicht, B. (2015). Area efficient integrated gate drivers based on high-voltage charge storing. *IEEE Journal of Solid-State Circuits, 50*(7), 1550–1559. ISSN: 0018-9200. https://doi.org/10.1109/JSSC.2015.2410797
2. Xu, J., Sheng, L., & Dong, X. (2012, September). A novel high speed and high current FET driver with floating ground and integrated charge pump. In *Energy Conversion Congress and Exposition (ECCE), 2012 IEEE* (pp. 2604–2609). https://doi.org/10.1109/ECCE.2012.6342393
3. Txapartegi, M., & Liao, J. (2017, March). *Gate driver market and technology trends.* Yole Développment. https://www.i-micronews.com/category-listing/product/gate-driver-market-and-technology-trends-2017.html#description
4. Seidel, A., & Wicht, B. (2018). Integrated gate drivers based on high-voltage energy storing for GaN transistors. *IEEE Journal of Solid-State Circuits, 1*–9. ISSN: 0018-9200. https://doi.org/10.1109/JSSC.2018.2866948
5. de Vries, I. D. (2002, March). A resonant power MOSFET/IGBT gate driver. In *Proceedings of the APEC Seventeenth Annual IEEE Applied Power Electronics Conference and Exposition (Cat. No.02CH37335)* (Vol. 1, pp. 179–185). https://doi.org/10.1109/APEC.2002.989245
6. Mashhadi, I. A., Ovaysi, E., Adib, E., & Farzanehfard, H. (2016). A novel current-source gate driver for ultra-low-voltage applications. *IEEE Transactions on Industrial Electronics, 63*(8), 4796–4804. ISSN: 0278-0046. https://doi.org/10.1109/TIE.2016.2554539
7. Mashhadi, I. A., Khorasani, R. R., Adib, E., & Farzanehfard, H. (2017). A discontinuous current-source gate driver with gate voltage boosting capability. *IEEE Transactions on Industrial Electronics, 64*(7), 5333–5341. ISSN: 0278-0046. https://doi.org/10.1109/TIE20172674626
8. Mashhadi, I. A., Soleymani, B., Adib, E., & Farzanehfard, H. (2018). A dual-switch discontinuous current-source gate driver for a narrow on-time buck converter. *IEEE Transactions on Power Electronics, 33*(5), 4215–4223. ISSN: 0885-8993. https://doi.org/10.1109/TPEL.2017.2723240
9. Kaufmann, M., Seidel, A., & Wicht, B. (2020, March). Long, short, monolithic – The gate loop challenge for GaN drivers: Invited paper. In *Proceedings of IEEE Custom Integrated Circuits Conference (CICC)* (pp. 1–5). https://doi.org/10.1109/CICC48029.2020.9075937

10. van Schuylenbergh, K., & Puers, R. (2009). *Inductive powering basic theory and application to biomedical systems*. Berlin Heidelberg: Springer Science & Business Media. ISBN: 978-9-048-12412-1.
11. Roschatt, P. M., McMahon, R. A., & Pickering, S. (2015, June). Investigation of dead-time behaviour in GaN DC-DC buck converter with a negative gate voltage. In *Proceedings of 9th International Conference on Power Electronics and ECCE Asia (ICPE-ECCE Asia)* (pp. 1047–1052). https://doi.org/10.1109/ICPE.2015.7167910

Chapter 7
Conclusion

There is an ongoing trend towards highly compact and efficient power electronics in fields like renewable energy, e-mobility, and industry. This is driven by the aim of lowering weight and volume, especially of portable power electronics, being cost-effective, and coping with space restrictions in more and more complex systems. Power transistors are key components in power electronics, achieving compact solutions by fast and efficient switching. Silicon power transistors and especially transistors made of wide-bandgap materials, like gallium nitride (GaN), offer excellent parameters and demonstrate significant progress over the years. To leverage their potential, gate drivers play an important role by supporting fast and efficient transistor switching, a high integration level, and flexibility regarding assembly.

This work focuses on gate drivers for power transistors in applications in the sub-10 kW-range, switching transistors at ≤ 400 V and currents ≤ 10 A, which is a typical range for fast-switching silicon and GaN devices. However, the solutions presented in this work are widely adaptable for various applications.

The fundamentals show that a low inductive connection between gate driver and power transistor is crucial to achieve fast switching. By integrating the buffer capacitor of the gate driver supply, a design with low parasitic inductance is facilitated. Furthermore, reliability is increased due to fewer connections, and the number of external components is reduced. However, the conventional method of integrating buffer capacitors suffers from a low gate charge delivery capability, leading to relatively large capacitor values, which cannot be integrated or which require large chip area, resulting in higher costs. A concept comparison of different driver output stages reveals that the driver output stage with high output current capability in a gate driver takes dominant chip area. Even the buffer capacitor (bootstrap capacitor) buffering the gate charge for the pull-up NMOS transistor of the gate driver output stage can hardly be integrated. Therefore, conventional driver output stages typically avoid bootstrap capacitors, but require relatively large high-voltage PMOS transistors. A detailed comparison of power transistors is provided

© The Author(s), under exclusive license to Springer Nature Switzerland AG 2021
A. Seidel, B. Wicht, *Highly Integrated Gate Drivers for Si and GaN Power Transistors*, https://doi.org/10.1007/978-3-030-68940-7_7

for silicon and GaN devices. Both switch types have their advantages depending on the application although GaN transistors are gaining more and more importance and market share. From the characteristics of the power transistors, requirements for the GaN gate drivers of this work are derived such as a three-level gate drive scheme to enable robust and efficient GaN transistor switching. Different driver supply and control concepts are presented and challenges and solutions are mentioned.

An area-efficient bootstrap circuit is presented, with fully integrated bootstrap capacitor that buffers the charge for the pull-up NMOS transistor of the gate driver output stage. The circuit incorporates the concept of high-voltage charge storing (HVCS) [1], which is developed and proposed as part of this work. HVCS utilizes a high-voltage capacitor that allows a higher voltage drop while discharging to the transistor's gate. Following the relation $Q = \Delta V \cdot C$, it significantly increases the charge delivery capability. In consequence, the bootstrap capacitors can be fully integrated on-chip, while a conventional bootstrap circuit would require an external discrete capacitor. An existing bootstrap circuit can be easily modified into the proposed solution by placing the new circuit portions in parallel. Three circuit options of the bootstrap circuit are presented. Besides gate drivers, the concept can be applied to most circuits using bootstrapping, such as integrated class D output stages or switched mode power supplies. To enable a permanently turned-on gate driver output stage, various charge pump circuits are compared to what extent they are suitable for implementation in the proposed bootstrap circuit. A calculation guideline for optimum sizing of both bootstrap capacitors is given considering worst-case corners to prevent a voltage overshoot as well as a too large voltage dip at the bootstrap capacitor. The most advantageous circuit option was manufactured in a 0.18 μm bipolar CMOS DMOS (BCD) technology. A high-voltage metal-metal capacitor and a poly-nwell capacitor are implemented on top of each other as bootstrap capacitors. They require 70 % less area than a conventional bootstrap capacitor with comparable charge delivery capability. The measured voltage dip close to 1 V confirms the calculations of the sizing guideline. The proposed area-efficient bootstrap circuit supports the implementation of compact and highly integrated power management systems. The concept of HVCS is very suitable to integrate the buffer capacitor of a gate driver for small or medium size GaN transistors. This contributes to the trend of monolithically integrated gate drivers together with the GaN transistor on the same die [2–11].

Fully integrated gate driver ICs for even larger GaN transistors are investigated. The drivers are based on the concept of high-voltage energy storing (HVES) [12], which was developed in this work as an extension of HVCS. HVES delivers high-current pulses with a large amount of gate charge from an on-chip high-voltage capacitor in resonant operation. Implemented in a 0.18 μm BCD technology, HVES enables to deliver up to 11.6 nC gate charge, to drive typical high-voltage as well as low-voltage/high-current GaN types currently available on the market. The concept of HVES is compared in detail to a conventional buffer capacitor implementation circuit and to the concept of HVCS. For small gate charges, the concept of HVCS shows benefits due to its low complexity. However, depending on parameters, HVES has about 10 times higher gate charge delivery compared to a comparable conventional circuit and 20 % higher than a circuit based on the concept of HVCS. Moreover, HVES has a ~25 % higher gate drive speed capability in comparison to the conventional capacitor implementation methods even at a three times larger gate loop inductance. The concept of HVCS is almost as fast as the concept of HVES. The gate charge efficiency of HVES is 1.3 times higher than that of HVCS because of its resonant operation. In its basic implementation, a gate driver with HVES provides a unipolar gate voltage. Such a driver is applied twice, at the gate and the source of the GaN device, leading to a full-bridge architecture. This results in the proposed fully-integrated gate driver providing a positive, negative, and 0 V gate-source voltage. This enables a bipolar and a three-level gate drive voltage scheme to ensure robust switching as well as low power losses during reverse conduction of the GaN device. The driver is suitable for normally-off GaN power switches including gate-injection transistors (GIT), by delivering a limited and constant gate current during GaN transistor on-state. The driver is flexible to operate from galvanically isolated and non-isolated driver supplies due to an adapted HVES circuit, which utilizes the existing on-chip high-voltage capacitor to achieve charge pump behavior. This charge pump generates a negative voltage supply rail from a galvanically non-isolated driver supply for the bipolar and three-level gate drive operation. With all these advantages, gate drivers based on HVES enable to better utilize the fast switching capabilities of GaN for advanced and compact power electronics.

Gate drivers are typically placed as close as possible to the power transistor to achieve the lowest possible gate loop inductance, enabling fast and robust switching. However, in many applications, this is not possible. Also, separation of gate driver and power transistor may result in better utilization of the power transistor and increased flexibility in packaging and assembly. The presented gate driver concept is based on the concept of HVES, which provides a higher gate charge speed at a given gate loop inductance compared to other buffer capacitor implementation concepts [13]. The HVES circuit of this driver utilizes the parasitic gate loop inductance for resonant behavior, avoiding an additional integrated inductor. A discrete capacitor connected in parallel to the gate-source terminals of the power transistor provides stability of the gate-source voltage of the GaN device. Experimental results show that the gate could be charged up to 40 % faster compared to a conventional gate driver. An improved conventional driver with a clamping circuit in parallel to the gate-source terminals is still 15 % slower than the presented driver. Transient measurements at a setup with the gate driver placed at a distance of ~25 cm from the GaN transistor show drain-source voltage slopes of up to 60 V ns^{-1}, at amplitudes of 400 V and 10 A. This gate driver for large gate loops has the potential for even faster switching.

Within the scope of this work, approaches and ideas for future work are developed. These include circuit modifications to achieve an even more compact gate driver output stage, more examples for applying the proposed capacitor implementation concepts (HVCS, HVES) to other circuits, and the extension of the circuits to achieve advanced functionality such as active gate control. Furthermore, another implementation option of the gate driver for large gate loops is presented that achieves the fast gate drive speed of HVES also at low gate driver supply voltages and promises even better gate driver efficiency.

References

1. Seidel, A., Costa, M. S., Joos, J., & Wicht, B. (2015). Area efficient integrated gate drivers based on high-voltage charge storing. *IEEE Journal of Solid-State Circuits, 50*(7), 1550–1559. ISSN: 0018-9200. https://doi.org/10.1109/JSSC.2015.2410797
2. Fichtenbaum, N., Giandalia, M., Sharma, S., & Zhang, J. (2017). Half-bridge GaN power ICs: Performance and application. *IEEE Power Electronics Magazine, 4*(3), 33–40. Navitas driver. ISSN: 2329-9207. https://doi.org/10.1109/MPEL.2017.2719220
3. *LMG3410 600V 12-A Integrated GaN Power Stage* (2017, April). Texas Instruments Incorporated.
4. Semiconductor, Dialog. (2016). *DA8801 SmartGaN^{TM} Integrated 650V GaN Half Bridge Power IC*. Dialog Semiconductor. https://www.dialog-semiconductor.com/sites/default/files/da8801_smartgan_product_brief.pdf
5. Rose, M., Wen, Y., Fernandes, R., Van Otten, R., Bergveld, H. J., & Trescases, O. (2015, May). A GaN HEMT driver IC with programmable slew rate and monolithic negative gate-drive supply and digital current-mode control. In *Proceedings of IEEE 27th International Symposium on Power Semiconductor Devices IC's (ISPSD)* (pp. 361–364). https://doi.org/10.1109/ISPSD.2015.7123464
6. Moench, S., Costa, M., Barner, A., Kallfass, I., Reiner, R., Weiss, B., et al. (2015, November). Monolithic integrated quasi-normally-off gate driver and 600 V GaN-on-Si HEMT. In *Proceedings of IEEE 3rd Workshop Wide Bandgap Power Devices and Applications (WiPDA)* (pp. 92–97). https://doi.org/10.1109/WiPDA.2015.7369264
7. *EPC2112 – 200 V, 10 A Integrated Gate Driver eGaN® IC – Preliminary Datasheet* (2018, March). Efficient Power Conversion Corporation.
8. Moench, S., Reiner, R., Weiss, B., Waltereit, P., Quay, R., Kaden, T., et al. (2018, June). Towards highly-integrated high-voltage multi-MHz GaN-on-Si power ICs and modules. In *Proceedings of Renewable Energy and Energy Management PCIM Europe 2018; International Exhibition and Conference for Power Electronics, Intelligent Motion* (pp. 1–8).
9. Zhu, M., & Matioli, E. (2018, May). Monolithic integration of GaN-based NMOS digital logic gate circuits with E-mode power GaN MOSHEMTs. In *Proceedings of IEEE 30th International Symposium on Power Semiconductor Devices and ICs (ISPSD)* (pp. 236–239). https://doi.org/10.1109/ISPSD.2018.8393646
10. Ujita, S., Kinoshita, Y., Umeda, H., Morita, T., Kaibara, K., Tamura, S., et al. (2016, June). A fully integrated GaN-based power IC including gate drivers for high-efficiency DC-DC converters. In *Proceedings of IEEE Symposium on VLSI Circuits (VLSI-Circuits)* (pp. 1–2). https://doi.org/10.1109/VLSIC.2016.7573496
11. Kaufmann, M., Lueders, M., Kaya, C., & Wicht, B. (2020). 18.2 A monolithic E-mode GaN 15W 400V offline self-supplied hysteretic buck converter with 95.6% efficiency. In *Proceedings of IEEE International Solid-State Circuits Conference – (ISSCC)* (pp. 288–290).
12. Seidel, A., & Wicht, B. (2018). Integrated gate drivers based on high-voltage energy storing for GaN transistors. *IEEE Journal of Solid-State Circuits*, 1–9. ISSN: 0018-9200. https://doi.org/10.1109/JSSC.2018.2866948
13. Kaufmann, M., Seidel, A., & Wicht, B. (2020, March). Long, short, monolithic – The gate loop challenge for GaN drivers: Invited paper. In *Proceedings of IEEE Custom Integrated Circuits Conference (CICC)* (pp. 1–5). https://doi.org/10.1109/CICC48029.2020.9075937

Index

© The Author(s), under exclusive license to Springer Nature Switzerland AG 2021
A. Seidel, B. Wicht, *Highly Integrated Gate Drivers for Si and GaN Power
Transistors*, https://doi.org/10.1007/978-3-030-68940-7

Printed in the United States
by Baker & Taylor Publisher Services